Robotics, Maintenance Programs, Engineering, and Skilled Trades
Essential Skills For Manufacturers

Louis Bevoc and Nathan Brusselli

Published by
NutriNiche System LLC

Louis Bevoc books...simple explanations of complex subjects

Robotics	3
Maintenance Programs	18
Engineering	35
Skilled Trades	76

Robotics
In Manufacturing

Nathan Brusselli and Louis Bevoc

Published by
NutriNiche System LLC

Louis Bevoc books...simple explanations of complex subjects

Introduction 5
History 5
Advantages 8
    Complexity 8
    Health and safety 8
    Continuous productivity 11
    Job creation and preservation 12
Disadvantages 13
    Unexpected deviations 13
    Lack of job skills 13
    Installation, maintenance, and upgrades 14
    Planned job loss 14
Future 14
    Detecting safety issues 15
    Programming themselves 15
    Communicating with the supply chain 15
    They like the way things are goings 16
    They are not willing to sacrifice jobs 16
    They do not want to invest the money 16
Summary 17

Introduction

Robots are more popular today than they have ever been in the past...and that popularity shows no signs of slowing down. In fact, robot usage will continue to grow well into the future because, quite simply, people want things easier, faster, and cheaper. Like it or not, robot technology is here for the long haul.

Robot technology has benefitted many different types of organizations, including those in the military, law enforcement, retail, and transportation, but it is particularly useful for companies that manufacture products. Manufacturers are all about assembly, and assembly is what robots do well. These machines are able to perform the same job over and over again without interruption. Since higher production output typically equates to higher profits, it is rather easy to see why robots are so valued by leaders of manufacturing facilities.

Some people see the field of robotics as progression, while others view it as a setback. The reasons for these two viewpoints will be discussed in the advantages and disadvantages sections of this book. However, before entering into that discussion, it is important to briefly explore the history of robotics to get a basic understanding of where it came from and how it evolved. The next section provides a glimpse of how this field came into being and grew to dominate certain aspects of business.

History

It is hard to pinpoint exactly when robotics first appeared, but some people believe that robots have origins that date back thousands of years when mechanics first came into use by humans. This belief might or might not be true, but this book considers the history robots to start in the early 1800s during the Industrial Revolution when machines began to make jobs faster and easier for people. It was during this period that steam and machine tools changed the way manufacturers produced products by facilitating and streamlining processes. However, it took until the early 20th century to think about machines as replacements for humans...and robotics technology has continued to evolve since that time.

The field of robotics is now part of the mainstream, but its popularity likely started in military operations. During World War I, remote-controlled robotic devices were used by the United States in an effort to preserve human lives. These devices had no fear and knew no boundaries as long as they were operational. Few people know that Nikola Tesla, an American immigrant, introduced a radio controlled boat used by the allies during WWI. Tesla was a great inventor whose electrical prowess was overshadowed by Thomas Edison, but that story must be reserved for another discussion in a different book. However, it can be said with confidence that Tesla was a pioneer in the development of robotics as we know it today.

The term "robot" was not referred to until the mid-1920s. History credits the coining of this term to a Czechoslovakian playwright named Karel Capek. Capek used robots to replace humans for the things they do not like to do. However, this concept was for entertainment purposes only and use of the word as we know it today was not until Sci-Fi books and films appeared around 1960.

The idea of robots in the early 1960s captured the imaginations of the public, and it was around this same time that robotics became reality. George Devol was a continuously thinking entrepreneur who invented the first robot. He sold his product, which he called Unimate, to General Motors. One might think that the world watched anxiously as the first known robot went to work in the Ewing Township, New Jersey assembly plant, but this was simply not the case. However, this robot likely saved some employees from injury or death because it performed a dangerous job that most employees feared.

In the mid-1960s, Stanford University started experimenting with robots. They purchased the rights to Devol's invention and expanded on it. Eventually, Stanford built the first robot capable of reasoning, and this opened the door to a wide variety of new robotic inventions that have continuously evolved.

Robots built in the United States in the early 1970s were often designed for war-related usages. This is not surprising considering the country was at the end of the very unpopular Viet Nam War. The death and destruction of this war were, for the first time in history, shown on television, and this lead to Americans realizing the need to protect people from dangerous situations. Laser guided weapons were put into use via robotics, but these weapons still required humans to operate them so their success was limited.

The first humanoid robots were developed by the Japanese in the mid-1970s. These machines were intelligent, mobile, and somewhat agile. For the first time ever, small parts and pieces could be picked up and moved and communication between man and robot was possible. Around the same time, a robot was developed by David Silver that functioned similarly to a human hand...including being able to respond to touching something via built-in pressure sensors. Silver's invention truly was awe-inspiring and, in the late 1970s, it was introduced to some large manufacturers for use in their assembly operations.

In the 1980s, things got a little crazy with robots. They were built to play musical instruments, compete in games, and interact with humans. It was also during this time that robotic technology was seriously considered for personal use. People proposed ideas that would make robots smaller and more affordable so consumers could use them in the comfort of their homes. Robots were not yet available as personal assistants, but they were headed in that direction.

The 1990s saw robots being used for surgical procedures. By this time, most people understood the value of robots, but their entry into medicine increased that value. Robots were able to work on sensitive areas such as the brain and spine due to the accuracy and precision they provided. In short, doctors and machines worked together to do some of the most advanced surgery ever seen and patients benefitted immensely.

The beginning of this century has seen robotics come into virtually every aspect of business. However, it has particular application in manufacturing where the need for repetitive movement is paramount. In some manufacturing plants, robots work with employees to build products. In other plants, robots function alone....sometimes doing the work of several people.

Robotic technology is used in manufacturing plants today to lower costs, raise quality, and increase productivity. Robots appear to be a godsend for production faculties, but there are some people who believe these mechanical wonders are not in the best interests of the manufacturing sector because they eliminate jobs cause other problems in workplaces. This book explores the benefits and challenges of robotics in manufacturing. It does not take a stance on the value of this technology, but it does provide insight into the known advantages and disadvantages so readers can decide for themselves. Let's start with the advantages section.

Advantages

Robot technology is continually evolving, advancing, and moving on to new frontiers. It has now touched virtually every industry in the world, including those with production as their primary purpose. Below are some advantages of using robots in manufacturing facilities.

Complexity

Robots can handle tasks that are too complex for people. For example, their dexterity and memory allow them to multi-task with no problems. They also do well with tasks that demand precision and speed such as applying adhesives to computer components or sealing electrical encasements for water resistance. Their superior vision and sensory abilities that allow them to perform job tasks with precision and accuracy that was previously unknown in manufacturing facilities.

Additionally, robots rarely make errors after they have been properly programmed. Humans tend to make more mistakes as the complexity of the tasks they are performing increases, but this is not the case for robots. They work quickly and accurately with very few errors and the complexity of the work has little or no effect on their performance.

In short, robots can handle very complex tasks without the fear of defective workmanship. These attributes make them very valuable to manufacturing operations that need accuracy to protect their customers. Pharmaceutical producers and medical equipment assemblers are examples of manufacturing operations that could jeopardize the lives of their customers if errors are made, and robots work well in both of these types of environments.

Health and safety

This benefit refers to the physical well-being of employees. Robots are an excellent choice for performing dangerous jobs. In manufacturing, these jobs might involve loading a press, laboring in extreme temperatures, or working around sharp blades. Mundane tasks can also be dangerous as employees become bored and their minds wander off to something else. For example, a worker who works all day loading cameras from a conveyor into boxes might start to daydream and end up pinching their fingers in the conveyor...a mistake that a programmed robot would not make.

One of the more common problems involving health and safety involves jobs that require repetition. Employees performing repetitive tasks can end up with carpal tunnel of the wrists, torn rotary cuffs, strained backs, vision problems, and other health issues. Personal protection equipment is available to prevent some of these injuries, but it is not always supplied or worn; thereby resulting in workers ending up in pain.

Interestingly, some repetitive injuries can occur in the offices of manufacturing plants rather than on the production floors. Continuous typing, sitting in the same position for hours, and starring at a computer screen for extended periods of time have all been known to cause injuries. These types of injuries show that repetition, in any form, can result in people getting hurt at work.

Some people think injuries from repetitive movement are few and far between and nothing to worry about. Unfortunately, this thinking could not be farther from the truth. Repetitious injuries cost employers millions of dollars every year, and can lead to employees becoming financially strapped, angry, depressed, and, in more severe cases, never returning to the workforce. These injuries do not happen to robots because they are programmed to do repetitive movement without ever stopping. If machine parts wear out, then they are simply replaced with no "healing time" required.

Other health and safety issue that can be prevented in manufacturing plants by replacing humans with robots include those listed below.

Physical violence

Workplace violence is a serious issue that is being discussed by leaders of organizations all over the world. It is a rather complex topic that cannot be fully described in the scope of this book, but it usually involves employees physically attacking each other. It can stem from just about any type of disagreement, and it can be planned or spontaneous. An example of planned physical violence is a woman who gets demoted and becomes upset as she thinks about it at home. She brings a gun to work the next day and, upon arrival, shoots the supervisor who is responsible for her demotion. An example of spontaneous physical violence is a situation where two workers getting into a disagreement over how much work each is doing. The argument becomes heated, and one of the workers physically assaults the other out of anger or frustration.

Regardless of how workplace violence occurs, it must be prevented or people risk being injured or killed. Robots are a good form of prevention because they do not get mad at humans or other robots. They complete their job functions while minding their own business, and they do not change unless programmed to do so.

Verbal attacks

Workplace aggression is not always physical. In fact, employees are injured more frequently by verbal attacks than they are by physical attacks. These verbal attacks usually affect the mind, not the body, and they cause people to become demotivated, mad, afraid, or depressed.

Unfortunately, some people think it is funny when they attack others personally. In workplaces, employees who attack other employees often have an audience of coworkers present that they believe are being "entertained." People witnessing the verbally aggressive behavior might laugh, but usually, this laughter is a result of discomfort. Unfortunately, people being targeted by the verbal aggressiveness find absolutely no humor in the situation.

If not prevented, verbal aggressiveness can lead to employees searching elsewhere for employment. When people leave organizations, their former employers lose the knowledge and experience of those individuals.

Robots are one potential solution to this issue because they eliminate the people who attack each other and prevent the psychological hurt that usually follows.

Workplace accidents

Accidents happen in many workplaces, but they are more likely to occur in manufacturing environments due to the equipment and machinery that is needed for production. Fingers and hair can get caught in machines, employees can fall when working at elevated levels, and slippage can occur in areas with wet or slick floors.

Robots prevent workplace accidents because they perform repetitive and exact movements that rarely deviate; thereby eliminating the employee mistakes that cause accidents to occur in manufacturing facilities. In the rare instances that robots malfunction, the worst that can happen is damage to machinery, and machine damage is always better than human damage.

Overexertion

Employees who overexert themselves are at risk of injury because they are pushing the limits of their bodies. They can strain muscles, tear ligaments, break bones, pass out, or have a stroke or heart attack. Overexertion is typically not life-threatening, but it a major cause of workplace injury and those injuries result in employee absenteeism.

Robots rarely injure themselves, they do not need time off from work to heal, and they show up or work every day. In terms of overexertion, robots are a much better choice than humans.

Continuous productivity

This is a major advantage of robots because they can essentially work 24/7 without a break. People in manufacturing understand the value of working around the clock in order to increase the productivity necessary to make more products. This equates to higher throughput, which means more business can be taken on by manufacturing plants. It also means a faster return on investment (ROI) because the added business offsets upfront costs.

Without a doubt, manufacturing leaders value continuous productivity in their plant operations. The increased amount of product that is generated opens the door to business partnerships that were not possible in the past. For this reason, interest in robotics is skyrocketing, and this infatuation is not likely to change anytime soon.

Job creation and preservation

One of the issues that people have with robots is that they eliminate jobs for workers. This might be true, but robots in manufacturing also create technical jobs that are more skilled and high paying. These jobs are sought after, long-term positions that grow in importance...unlike many general assembly positions in production facilities. In short, some jobs might be sacrificed due to the implementation of robotics, but better jobs evolve for the same reason.

Robots also help protect jobs due to the fierce completion in the global marketplace. Companies that cannot compete use robotics to narrow the gap. When this happens, work is not outsourced to other nations and existing jobs are protected. Like it or not, robotics are an important aspect of manufacturing...and they help create and preserve jobs.

Some people would argue that robots do nothing for job creation and preservation. However, one only needs to look at the facts to realize that their arguments have little, if any, validity. Quite simply, North American manufacturers that cannot compete due to high labor costs are going to outsource their work to companies that pay lower wages. Those companies are typically located in countries outside of North America; thereby resulting in North American jobs being lost. Outside of wage and benefit concessions, which people rarely accept without a fight, robots are often the only way to prevent the loss of these jobs.

As might be expected, the use of robots in manufacturing is not thought to be beneficial in the minds of certain people. Some of the reasons these individuals feel the way they do are discussed in the next section.

Disadvantages

Robot technology has generated interest of most business leaders. However, it must be noted that the results have not all been positive. Below are some disadvantages of using robots in manufacturing facilities.

Unexpected deviations

As noted earlier, robots do a great job performing the things that are expected of them. Once programmed, they are able to complete detailed assembly tasks over and over for unspecified periods of time. Quite simply, they do what they are supposed to do without stopping, arguing, or complaining about the workload.

Robots might do well with the expected, but they have problems when they encounter unexpected deviations. They cannot change without being programmed to do so. In other words, they cannot react to situations that require change on the spot. If change is required, then production has to stop for reprogramming. Service of this type is counterproductive, and the downtime costs quickly add up. If reprogramming continues to be necessary, then the value of the robot begins to diminish.

Lack of job skills

Robots are often the epitome of technology. As such, they usually need skilled technicians to program, operate, maintain, and oversee them. People who have all of these skills are usually paid quite well, but they are difficult to find and limited in number. Their uniqueness makes them highly marketable, and manufacturers need to compete to hire them. If they cannot be found and hired, then the robots will not perform as expected, and productivity will suffer.

Another aspect of job skills that is often overlooked is training. Whoever possesses the program, operation, and maintenance skills must be able to show others so they can acquire those same skills. Training others can be a challenging and arduous task, and many people simply are not able to do it. They lack the people skills, patience, desire, or ability to transfer their knowledge in a way that is understood by the trainees. Regardless of the reason why, a skilled person's inability to train other employees means the robots will not operate as expected and might even become more of a liability than an asset.

Installation, maintenance, and upgrades

It has been said that money does not buy happiness. This statement is likely true, but it can also be said that lack of money often creates misery...especially when manufacturers need to install, maintain, and upgrade machinery. These costs tend to escalate when robotics are involved, but the general thinking is that the resulting increased productivity will exceed those costs and open the door to new markets for sales and distribution. However, some manufacturing leaders cannot bring themselves to invest the money required to make it happen.

Planned job loss

This disadvantaged cannot be overlooked, even though it is was challenged in the advantages section of this book. Yes, robot technology does create high paying technical jobs and it prevents manufacturing jobs from leaving the shores of North America. However, regardless of the effort made, some jobs will be lost in the process. After all, one of the goals of robotics is to reduce costs, and costs are reduced when jobs are eliminated.

For some people, job loss alone is enough to discourage robot technology in manufacturing plants. Many of the employees who lose their jobs are unskilled production workers who have difficulty finding comparable work. Some do not have reliable transportation and others are living paycheck to paycheck due to the relatively low wages that they are typically paid. Technology is not necessarily a reason for production workers to panic, but robots add to the economic disadvantage they are already experiencing, and this makes their financial existence even more difficult. In short, planned job loss boils down to choices...and those choices involve morals, ethics, and societal obligations.

Future

This is probably the most interesting section of this book since the use of robots in manufacturing is virtually unlimited. They can assemble, scan, inventory, stack, glue, press, seal, and weld parts or finished products with little or no assistance; thereby increasing their popularity in production facilities. Based on the capabilities and wide range of applications, the future of robotics appears to be very bright. In fact, robot technology has little threat of going backward, and manufacturers will be one of the major beneficiaries.

The presence of robots will continue to grow in all businesses, and they will play more critical roles in the manufacturing sector…from raw materials to finished products. Future uses for robots in production facilities include:

- *Detecting safety issues*

    Technological advances are valuable because they usually bring about positive change in efficiency and effectiveness. This is true for most aspects of manufacturing, and it is certainly true for robotics. However, one area unrelated to efficiency and effectiveness where robot technology has great potential is safety. More specifically, robots will be able to detect and alert employees to unsafe areas in workplaces using artificial intelligence; thereby preventing humans from getting injured. The pain and suffering avoided and money saved from this advancement will make even more manufacturers jump on the robot bandwagon.

- *Programming themselves*

    As crazy as this sounds, self-programming is something that will occur in the future. This type of technology is already being seen in self-driving automobiles, and it will be expanded upon in manufacturing facilities. When this happens, employees will be able to gather information without being trained by a limited number of competent individuals. Robots will be the trainers and their knowledge base will never be lost.

- *Communicating with the supply chain*

Software and scanning technology will be available for robots to talk to other robots throughout the supply chain. When this happens, accurate information will be transferred from one area of the supply chain to another at lightning speed without the need for people. Issues that hinder human communication such as time zones differences, health issues, personality clashes, stress, and absenteeism will no longer factor into the process. In short, robot communication will facilitate the supply chain process by speeding it up and reducing communication errors.

Regardless of the advantages offered by robotics, some business leaders simply cannot or will not bring their production facilities into the technological age. Reasons for their resistance include:

- *They like the way things are going*

    Some leaders believe in running their operations in ways that have worked in the past. Manufacturing leaders who think like this way usually adhere to the "if it ain't broke, don't fix it" mentality. They are comfortable with their manufacturing operations and have no plans to make changes just because the technology is available. In some ways, this makes sense because the implementation of robotics involves venturing into the unknown. However, in future, the amount of "unknown" will continually decrease…and more leaders will buy into the positives of this change.

- *They are not willing to sacrifice jobs*

    This type of resistance is difficult to argue because it takes the general welfare of people into consideration. Manufacturing leaders who have compassion for employees and their families might not want to eliminate jobs, regardless of how few will be lost in the process of the robotics implementation. However, like it or not, leaders who fight progress to save jobs will be fewer and farther between in the future. Some will adhere to their moral compass, but others will see the value of robotic technology and go with the flow.

- *They do not want to invest the money*

The biggest challenge to the bright future of robotics in manufacturing is cost. More specifically, the cost at startup can be quite expensive and prohibitive. In fact, it can be so expensive that the ROI seems rather unachievable, so the idea of robotic technology takes a back seat to traditional methods.

It can be said with confidence that most manufacturing leaders would not have achieved their positions if they were completely resistant to change, but some operate with a little bit of this mentality...especially if that change has a large upfront cost with no guarantee of recovery. However, like all technology, robotic costs will go down as their capabilities go up...and this will make it much easier for leaders to part with initial investments.

Summary

Robotics has evolved as a field quite rapidly over the past century, and this evolution shows no sign of being in the final stages. In fact, robots are being used in virtually every type of industry, business, or organization that exists today. They lighten workloads, make better use of time, speed up activities, and ease the lives of people all over the world. They work well in quality and quantity applications, and many people believe we have only seen the tip of the ice burg in terms of their value to humans.

This book explores robotics in manufacturing. It examines the progressive history and evolution of robots while highlighting the advantages and disadvantages that that fuel debate over their effectiveness and value. It also discusses the future of robotic technology while suggesting ways that it could lessen existing burdens in production facilities. The text is educational and informational, and it is written for easy understanding at all reader levels.

Congratulations! You now understand more about robotics in manufacturing...a rapidly growing requirement in production facilities worldwide.

Maintenance Programs in Manufacturing
An Introduction to Preventative, Predictive, and Corrective Types

Louis Bevoc

Published by
NutriNiche System LLC

Introduction	20
Planning	22
Who will be involved?	23
What machinery will be involved?	23
What are the priorities?	23
What will be documented?	23
What is the operating budget?	24
Preventative maintenance	25
Advantages	25
Disadvantages	26
Predictive maintenance	28
Advantages	28
Disadvantages	29
Corrective maintenance	30
Advantages	31
Disadvantages	32
Summary of the three types	33
Summary	34

Introduction

Welcome to a world that is vastly underestimated in value. It is a world where "good morning" is replaced with "I have a machine down." It is a world where employees work before, during, and after production activities. It is a world where criticism is frequent and praise is rare. It is the world known as maintenance in manufacturing.

The above paragraph might be a little dramatic, but it is reality in many manufacturing facilities. Maintenance keeps production lines operating so orders can be filled and customers will be happy. In this regard, maintenance is one of the most important aspects of a production facility. That being said, what are the specific duties of maintenance personnel? The following is a list of their major job responsibilities:

Assembly and disassembly

When new machinery enters the facility, maintenance people are the first to have contact with it. They uncrate it, assemble it, and set it up. They are responsible for making sure all parts are included and the machine will do what it is supposed to do.

Maintenance people also disassemble machinery when it needs to be broken down for cleaning, repair, or removal. They understand the inner workings of the machines, and they know how to properly dismantle them without damaging them or creating safety risks for themselves or others.

In short, the buck stops with maintenance people in terms of assembly and disassembly.

Machine repair

Machines break down in manufacturing facilities. This happened in the past, it happens now, and it will happen in the future. However, broken equipment is not a major issue if it can be fixed...and maintenance people fill the role of fixers. They understand what needs to be done in order to get machines running properly in a reasonable amount of time. This saves manufacturers money, and it allows them to get their products to the customers who need them. In terms or machine repair, maintenance people are worth their weight in gold.

In short, machines need to be repaired and maintenance people fill that need.

Liaison

When machinery cannot be repaired, it needs to be serviced by external professionals. This means a technical person from the manufacturer of the machine needs to visit the facility to make the necessary repairs. Someone needs to be responsible for contacting that technical person and describing the problems the machine is experiencing...and that someone is almost always a maintenance person.

Describing machinery problems might seem like a relatively simple task. After all, the machine is broken, so what else needs to be said? Unfortunately, a lot more usually needs to be said. Technical people need detailed information in order to make a timely and proper diagnosis, and that information is only available from those who have a working knowledge of the machine. Typically, the only individuals with that working knowledge are maintenance people.

In short, maintenance people are liaisons who reduce the time, effort, and expense required for external repair of machinery.

Testing

How do production people know if a machine is able to meet their expectations? The answer is through testing. For example, a light intensity machine at a flashlight manufacturing company needs to measure the bulb brightness of products made on a high-speed production line. One bulb needs to be measured every six seconds in order for plant personnel to meet established production quotas. Every day before production begins; maintenance people run tests to make the light intensity machine can handle the volume.

In short, maintenance people verify machines are capable of doing what they are supposed to be doing.

Calibration

How do production people know that a machine is working properly? The answer is through calibration. For example, a scale in a meat processing plant weights weighs one pound packages of hot dogs. It is definitely capable of weighing these hot dogs, but is it producing accurate results? The only way to find out is by calibrating the scale with standard weights...and this is done by maintenance people.

In short, maintenance people verify machines are accurately doing what they are supposed to be doing.

This book focuses on the three major types of maintenance programs in manufacturing facilities known as preventative, predictive, and preventative maintenance. It examines the advantages and disadvantages of these programs in layman's terms. Maintenance terminology can be quite complex, but the text in this book is written so it is easily understandable at any reader level.

Now that you understand the scope of this book, we can into a discussion on the three major types of maintenance programs. However, before doing this, we need to discuss the planning of these programs in order to get a better understanding of how they are implemented.

Planning

Maintenance is very important in manufacturing facilities because it affects the livelihood of everyone in the organization. Machines need to run properly because broken machinery stops production...and production is the life-blood of manufacturing operations. In fact, production is a major reason that most manufacturing plants exist. Based on this fact, it is relatively easy to understand the importance of maintaining machines in proper working order.

Maintenance programs need to be planned before they are implemented...regardless of whether the type of maintenance is preventative, predictive, or corrective. Planning starts by defining a purpose. Will the maintenance program be proactive, reactive, or projective? If the program is proactive, then the purpose is preventative maintenance. Machines will be serviced on a regular basis to prevent problems from occurring during production. If the program is projective, then the purpose is predictive maintenance. Machine failures will be predicted so they can be managed. If the program is reactive, then the purpose is corrective maintenance. Machines will be serviced when they break down.

Surprisingly, most manufacturers do not have predictive or preventative maintenance programs in place, choosing instead to take corrective action as needed. Money and time are two major factors that cause them to opt out of predictive or preventative activities because these resources are needed in other areas. This plan might work in the short term, but it can cause a wealth of machine problems down the road. However, regardless of the type of maintenance program chosen, the purpose of it needs to be defined.

Next, the scope of the program needs to be outlined. Questions that need to be answered include:

Who will be involved?

What people are going to perform the work and what are their designate responsibilities? In a manufacturing facility, the scope might only be applicable to one department or it might encompass the entire plant. Some maintenance personnel are highly skilled while others are not, so specific roles need to be defined. There also needs to be a manager in charge who reports to upper management.

What machinery will be involved?

Manufacturers need to designate the equipment or machinery that is going to be maintained or repaired. Some machines are purposely left off of this list because they are only serviced by representatives of the companies that manufacture them. Other machines might be left off this list because they are located outside of the physical boundaries of the plant. For example, company-owned vehicles (trucks, bulldozers, tractors, etc.) might not be serviced by maintenance personnel simply because it is more convenient and cost-effective to use external sources.

What are the priorities?

Does one machine or department take priority over others in terms of repair? This is an important question because it gives maintenance personnel guidelines for allocating their time. Without a list of priorities, time can easily be spent fixing machines that are not needed until a later time. The focus needs to be on equipment that is immediately needed for production.

What will be documented?

If utilized, will routine maintenance checks of machinery and equipment be recorded? Routine maintenance is done in many plants on a periodic basis, and it is nearly impossible to remember all of the dates and times that service was performed.

A document should be available that lists all routine maintenance checks and the frequency that those checks are performed. Frequency should be based on usage (volume), safety, and manufacturer recommendations. Documentation of routine maintenance serves three basic purposes:

> *It indicates when maintenance needs to be conducted*
>
>> Documentation shows when a machine was last serviced, and when the next service is due. This eliminates the need to rely on memory. Technology today even allows for reminders of upcoming services that can be delivered directly to smartphones or other mobile devices.
>
> *It maintains warranties*
>
>> Some warranties do not remain in effect if certain routine maintenance is not performed. For example, oil might need to be changed in a machine every 200 working hours or the warranty is not valid.
>
> *It provides information for authorities*
>
>> OSHA, auditors, and government agencies all request routine maintenance information when investigating injuries, validating processes, or probing violations. IF this information is not provided, fines could be levied and customers could be lost.

What is the operating budget?

Like every other aspect of a business, money plays a role in maintenance planning. Many manufacturers have budgets in place that limit expenses in maintenance departments. They allocate funding for various areas such as building and grounds upkeep, machinery repair, vehicle servicing, and building renovations.

Keep in mind that budgets are great for planning, but they do not always work for maintenance because there are often unforeseen circumstances. Machinery that is critical to production simply cannot wait until the next budget renewal...it has to be fixed now or the plant will not be able to operate and orders will not be filled.

Now you understand the importance of planning for maintenance programs. Armed with this knowledge, it is time to move into the specific types of programs...starting with preventative maintenance.

Preventative maintenance

In general, preventative maintenance programs are implemented so future problems can be avoided. For example, tires on a car should be rotated every 8000 miles. This prevents the tires from wearing unevenly and needing to be replaced before the 40,000-mile life expectancy. In a manufacturing facility, preventative maintenance is also designed to prevent future problems from occurring. For example, working machine parts need to be greased on a weekly basis to avoid excessive friction that leads to damage.

Like most other aspects of business, preventative maintenance has benefits and drawbacks in manufacturing plants. These positives and negatives must be taken into consideration when deciding whether or not to implement a preventative maintenance program. The time, effort, and money invested into this type of program needs to have a payback in order to be justified...and that justification can only be determined by plant management personnel.

The following are some specific advantages and disadvantages:

Advantages

Below are some advantages of preventative maintenance programs.

Risk reduction

Preventative maintenance reduces the risk that there will be failures during production. It provides insurance that production quotas will not be interfered with by broken machinery or faulty equipment. This eliminates a major headache for management personnel and allows them to focus on other areas of their jobs.

Life expectancy

Every manufacturer wants machinery and equipment to last as long as possible in their facilities. This eliminates replacement costs that can be quite significant...especially when machines are designed for a single purpose. That being said, machinery and equipment are expected to last longer when a preventative strategy is utilized because periodic servicing helps maintain them in proper working order. In terms of life expectancy, an ounce of prevention is truly worth a pound of cure.

Cost savings

When done properly, preventative maintenance easily justifies its existence economically. It helps (1) prevent maintenance personnel from fixing machines at a later date, (2) avert the need for external sources of repair, (3) avoid unnecessary downtime and the cost of unproductive labor, and (4) eliminate sluggish equipment that slows productivity. Based on these four areas of cost savings, it is rather obvious that the payback for preventative maintenance programs can be substantial.

Energy

This advantage goes largely unnoticed, but it needs to be noted. Machinery and equipment that are not serviced using preventive maintenance are often less energy efficient. They require more electricity or gas to function at desired levels, thereby increasing energy bills and wasting resources. Cost goes down and efficiency goes up when preventative maintenance programs are in place

Disadvantages

Below are some disadvantages of preventative maintenance programs.

Immediate costs

There is an up-front cost for preventative maintenance programs. Personnel need to be hired and supplies need to be inventoried. Depending on the number of people hired and the scope of the program, this can be quite expensive...and all of the costs are accrued before the first product leaves the production line. Some companies cannot afford to put out the money, and others simply refuse to do it because they believe the cost is not justified.

Management

This refers to maintenance people and preventative maintenance programs because both of them need to be managed. People require direction, and that directions needs to come from a supervisor. That supervisor also needs to make sure the program is followed and the work required gets done in a timely manner. Management of a preventative maintenance program is not a small task, and that makes it a disadvantage for manufacturers.

Volume changes

When is preventative maintenance too much or too little? This question is difficult to answer, and it can create a problem for manufacturers. For example, a program requires maintenance personnel to replace the wheels on all smokehouse racks in a turkey processing plant every six weeks. This makes sense because racks could break down during production, causing downtime. However, this program does not account for seasonal volume shifts such as Thanksgiving (when production is at a peak), and the summer months (when production is very low). In reality, wheel replacement should be much higher around Thanksgiving and much lower in the summer months...but this is not the case because the preventative maintenance program calls for replacement every six weeks.

Value

Unfortunately, some business leaders believe preventative maintenance is a luxury. As a luxury, it is one of the first areas to undergo cuts when manufacturers are experiencing financial difficulties. From a cost-saving perspective, this makes little sense because the money spent preventing problems is typically less than that spent repairing machinery or equipment that has failed. However, these leaders' thinking will likely never change because preventative maintenance is regarded as a "precautionary" expenditure that is difficult to tie to actual production downtime. If the value of something cannot be directly measured, then number crunchers in the manufacturing organization will push for its elimination during tough times.

Now you understand the advantages and disadvantages of preventative maintenance. This program is beneficial for many manufacturers because it keeps equipment and machinery operating at optimal levels. However, there are some up-front costs involved, and management needs to determine the real value of this program.

Next, let's move into a discussion on a type of program that uses logic and reasoning to assess maintenance needs. That program is known as predictive maintenance.

Predictive maintenance

This is the rarest type of maintenance program used by manufacturers. Essentially, machines and equipment are examined in order to predict when maintenance should be performed. Similar to preventative maintenance, predictive maintenance is implemented to avoid future problems. However, if done properly, this program costs less than preventative maintenance because service is only performed when justified. In other words, predictions are made about the potential failure of machines and service is performed just before those failures become reality. The goal is to avoid unnecessary maintenance expenses.

Predictive maintenance is also the most difficult type of maintenance program used by manufacturers. Timing is critical because service has to be performed before the failure with sufficient warning time must be provided. Techniques include observing machine performance, ultrasound, acoustics, thermal imaging, vibration analysis, and oil analysis.

The following are some specific advantages and disadvantages.

Advantages

Below are some advantages of predictive maintenance programs:

Maintenance time

This program predicts service needs. It falls under the preventative maintenance category, but service is only performed when it is justified by the potential for equipment or machine failure. This means less maintenance effort is needed for predictive maintenance, and the end result is a savings in time and labor.

Inventory

This is likely the least known advantage of predictive maintenance. Predictive maintenance does not require an inventory of excessive machine parts because only the parts necessary for the program are kept in stock. Emergency spare parts stock no longer exists, and this results in cost and space savings.

Safety

Skill levels of personnel performing predictive maintenance are high because these individuals have undergone training and understand the machines and equipment they are servicing. As a result of their knowledge, safety levels increase throughout the plant. This safety is critical because many manufacturers work with hazardous chemicals or operate equipment that requires high pressure or temperature. It creates a win-win situation for employees and management because employees do not go through the pain and suffering associated with injuries, and management does not pay the costs associated with workers compensation.

Disadvantages

Below are some disadvantages of predictive maintenance programs.

Monitoring/testing costs

Specialized monitoring and testing devices are typically expensive. They are made for a specific task, so a higher price can be charged for them. The cost might be understandable, but it is also prohibitive for some manufacturers. They either cannot or will not spend the money necessary for the equipment, so the predictive maintenance program does not function properly.

Required skills

Every employee is not capable of performing predictive maintenance tasks. In fact, the vast majority of employees are not capable of performing these tasks because they require specific skills. In addition to having mechanical skills, people who do predictive maintenance often need training in electronics, hydraulics, or thermodynamics.

Environmental effects

Some manufacturing plants have conditions that are less than ideal for the monitoring or testing devices necessary for predictive maintenance. For example, food processors with wet or cold working environments might have problems keeping these devices working properly. The same goes for the high temperatures found in foundries or smelting plants. Along the same lines, paint manufacturers are likely to have corrosive chemicals that could do damage.

Regardless of the way the damage is done, monitoring or testing devices that are not working properly will not provide accurate information. This means calculations could be inaccurate, and the entire predictive maintenance program is jeopardized. Since it is difficult for some manufacturers to avoid destructive work environments, it is understandable why they choose not to implement this type of program.

Now you understand some of the advantages and disadvantages of predictive maintenance. This program is beneficial for many manufacturers because it analyzes machinery and predicts when it will fail. This information is then used to perform service before the failure occurs while housing fewer parts and maintaining lower labor costs for maintenance personnel. However, people need specific skills to be employed as predictive maintenance technicians, and the type of work environment can affect the data collected.

Next, let's move into a discussion on a type of program that addresses machinery and equipment failures after they occur. That program is known as corrective maintenance.

Corrective maintenance

This is the most common type of maintenance program in manufacturing plants. It is a completely reactive program, and it is necessary because equipment and machines will break down at some point. Machines cannot be expected to run forever, and constant use at full capacity typically shortens that life span.

Unfortunately, corrective maintenance is often the only type of maintenance program available in a manufacturing facility….with no predictive or preventative programs to support it. Many times this is due to cost because smaller manufacturers cannot afford to sacrifice the resources necessary to set up predictive or preventive programs. However, sometimes corrective maintenance stands alone simply because organizations do not want to invest the necessary time and effort to establish other programs. Corrective maintenance does wonderful things for equipment and machinery repair, but it needs help. Without some type of support, corrective maintenance programs can become very expensive in a relatively short period of time. This adds stress to the jobs of maintenance personnel and managers, and they might start looking for jobs elsewhere.

Advantages

Below are some advantages of corrective maintenance programs.

Initial investment

Corrective maintenance does not require the planning, time, or effort required for preventative and predictive programs. This is because corrective maintenance does not address problems before they occur; it simply reacts to issues at the time of failure. This is advantageous for manufacturers because they save on resources. In short, there is a savings on initial investment for manufacturers that choose to corrective maintenance as their only maintenance program.

Expenses

Corrective maintenance delays expenses. These expenses include the services and checks performed under preventative and predictive maintenance programs. This means machines and equipment can function for extended periods of time with little or no maintenance. This strategy is particularly beneficial for manufacturers looking for short term return on investment, such as that expected from machinery or equipment needed for a specific purpose. For example, a toy manufacturing company might need a machine to stitch stuffed animals for two months until they implement an entirely new process. Management does not want to put time and money into maintenance of this machine unless it completely fails. Even if the machine breaks down, it will be "quick-fixed" or temporarily repaired because it will not be needed for the long term. In this case, the short term return justifies corrective maintenance being the only program in effect.

Profitability

Profit is a major reason that most manufacturers are in business, and higher profits can be made by organizations that prefer to react to maintenance issues as they occur. They are willing to forego using any type of predictive or preventative programs in order to make more money. Savings from labor, supplies, and parts all lead to higher profitability...and a happier management team.

Disadvantages

Below are some disadvantages of corrective maintenance programs.

Predictability

As most manufacturers are aware, this is likely the biggest disadvantage of a corrective maintenance program. Maintenance personnel do not know when equipment or machinery will fail, and that can cause a variety of different problems. For example, parts might need to be ordered, thereby delaying the repairs necessary to get production running. Additionally, outside service might need to be called in for issues that cannot be resolved by plant personnel...and that service is typically quite expensive. When these problems mount, the cost of the corrective maintenance program can far exceed that of a program that had preventative measures in place.

Efficiency

Efficiency is important for every production oriented facility because increasing it helps keep costs down and maximizes productivity. Unfortunately, corrective maintenance programs do little for efficiency. The major goal of corrective maintenance is to keep equipment and machinery operating, but optimal levels of operation are not necessarily part of that goal. When optimal levels are not achieved, equipment and machinery do not reach their potential...and the resulting lack of efficiency causes a decline in productivity.

Urgency

As has already been stated in this book, corrective maintenance programs do not prevent problems. This makes the likelihood of problems much more probable, and those problems need to be addressed immediately when they affect production. As might be expected, most machinery problems hinder production, so repairs need to be made with no time to waste. Unfortunately, this type of environment creates a wealth of stress for maintenance personnel and managers...and that is why urgency is a negative associated with corrective maintenance.

Now you understand some of the advantages and disadvantages of corrective maintenance. This program is beneficial for many manufacturers because it minimizes expenses and raises profitability. However, repairs necessary for corrective maintenance difficult to predict, and there always tends to be a sense of urgency.

The next section summarizes preventative, predictive, and corrective maintenance programs for a better understanding.

Summary of the three types

The three major types of maintenance programs have now been described. However, their specific applications might still be a little difficult to understand unless they are compared side-by-side. Based on this thinking, a brief and concise summary is as follows:

> Preventative maintenance programs are proactive, predictive maintenance programs are selectively proactive, and corrective maintenance programs are reactive.

Examples of the work performed by each type of program are as follows:

> *Preventative maintenance* – Greasing or lubricating working machine parts, changing oil in machinery
> *Predictive maintenance* – Measuring the amount of vibration on machines, searching for gas leaks in machinery
> *Corrective maintenance* – repairing machinery after it has broken down, replacing broken safety covers machines

In a nutshell:

> Preventative maintenance programs administer a wide variety of services that prevent failure, predictive maintenance programs do specific testing to determine services that prevent failure, and corrective maintenance programs perform services after a failure has occurred.

Based on what is written in this book, it is rather obvious that maintenance programs are necessary to keep equipment and machinery functioning properly in manufacturing facilities. That being said, existing maintenance programs need to be continually updated and improved upon. This might not be easy, but it is important...and it could affect the survival of some manufacturers.

Summary

Maintenance personnel are essential for any type of production-oriented operation. They assemble, diagnose, repair, and monitor the equipment and machinery needed to fill orders and satisfy customers. In terms of manufacturing, maintenance departments are the glue that holds facilities together.

This book focuses on maintenance programs in manufacturing. First, it examines the planning that takes place before maintenance programs are implemented while discussing the people, priorities, and documentation involved. Then it analyzes preventative, predictive, and corrective programs through description and an exploration of their advantages and disadvantages. The text is educational and informational, and it is written for easy reader understanding at all levels.

Congratulations! You now understand more about preventative, predictive, and corrective maintenance...three important types of maintenance programs used by manufacturers.

Engineering in Organizations
A Basic Introduction to the Mechanical, Electrical, Chemical, and Civil Branches

Nathan Brusselli and Louis Bevoc

Published by
NutriNiche System LLC

Mechanical Engineering	37
Electrical Engineering	48
Chemical Engineering	58
Civil Engineering	67

Mechanical Engineering

Introduction 38
Education 38
Natural ability 39
Responsibilities 41
Research and development 41
Equipment and machinery design 41
Equipment and machinery modification 42
Process design 42
Training 43
Project management 43
Work environment 43
Manufacturers 43
Job Shops 44
Laboratories 44
Offices 44
Travel 45
Temporary assignments 45
Future 45
General 45
Specific 47

Introduction

Engineering is the application of math and science to design, build, utilize, and maintain a wide variety of equipment, machinery, systems, and processes in organizations. It is a very broad field that can be broken down into many different subcategories that deal with specific aspects of the discipline. However, most experts break engineering down into four main branches for simplification purposes. These branches are mechanical engineering, electrical engineering, chemical engineering, and civil engineering.

Mechanical engineering concerns the construction and running of machinery and it will be the first branch discussed in this book. A mechanical engineer (ME) performs a variety of different tasks at work, but his or her major job function is defined as:

> The design, manufacturing, testing, maintenance, and usage of mechanical systems.

The roots of mechanical engineering can be traced back thousands of years....even to the actions of people in ancient Greece (BC). However, it was Sir Isaac Newton and his Laws of Motion that brought mechanical engineering in the spotlight using physics and mathematics as support. Years late, the first professional society of mechanical engineers was formed in Europe and a separate branch of engineering was officially developed.

Modern-day mechanical engineering involves physics, math, and science...not to mention good old fashioned common sense. People in this profession also have an understanding of thermodynamics, kinematics, mechanics, and electrical components. They utilize their knowledge to develop, test, and analyze many different types of machinery, equipment, processes, and systems in places such as manufacturing plants and research labs. Their work typically has an industrial application, and they are often involved with projects from conception to finish.

Top-notch MEs combine education with natural abilities for effectiveness. These qualifications are defined in more detail as follows:

Education

Most mechanical engineering jobs require a bachelor's degree in Mechanical Engineering. This is obtainable from many different universities, and it usually takes four to five years to complete. In the United States, this degree is accredited by the *Accreditation Board for Engineering and Technology* (ABET). The ABET regulates course requirements for all colleges and universities to ensure adequate training and equal standards, and approved programs are listed on their website.

Much of the coursework in bachelor's degree programs focuses on math and science related concepts. Math typically consists of higher level calculus and differential equations, and science courses revolve around physics...but chemistry is also significant. It is also important to note that the other three main branches of engineering (chemical, civil, and electrical) are part of a mechanical engineer's curriculum.

The following types of courses are typically taken be MEs in order to earn their bachelor's degrees:

- Calculus
- Chemistry
- Computer-aided design (CAD)
- Computer-aided manufacturing (CAM)
- Differential equations
- Engineering design
- Engineering technology
- Engineering theory
- Fluid dynamics
- Fluid mechanics
- Hydraulics
- Instrumentation
- Kinematics
- Linear algebra
- Machine design
- Physics
- Pneumatics

In addition to the course work, most degree programs require the completion of a project to graduate. This usually takes place in a business rather than an educational facility, and it is designed to provide students with real-world experience and let them apply their problem-solving skills.

Natural ability

Mechanical engineers' natural ability is often just as important as their education. In fact, MEs with highest natural ability typically turn out to be the most successful.

Natural ability includes:

Active listening

> This involves listening to what people are saying rather than just hearing them say it. There is a difference because hearing often goes "in one ear and out the other" while listening processes information for well-thought responses. Active listening is often a skill that is in high demand and short supply because people get distracted and fail to hear what is being said to them. This natural ability is important for many employees in organizations, but it is critical for MEs in order to ensure they have attained the correct information for problem-solving.

Creativity

> Some people think creativity is a natural ability that is only necessary for artists, musicians, poets, and writers. People in these professions need creativity…but so do engineers. This is due to the fact many problems are not easily resolved, and creativity is sometimes an important part of the solution. Mechanical engineers face a wide variety of machine related issues that need to be overcome, and creative ability helps them find workable answers to the problems they encounter.

Innovation

> Innovation is the process of coming up with new ideas or improving a concept that already exists. Based on this, it is understandable that innovation is a desirable quality in every mechanical engineer. After all, part of their job is to design and construct mechanical systems, and this requires original thinking. Without the natural ability to be innovative, MEs will not perform at levels expected of them by leaders of organizations.

Mechanical aptitude

> This is the most obvious natural ability that is important for MEs. Mechanical engineers have to understand how things work….especially the inner workings of machinery. Their mechanical aptitude helps them solve problems where formal

education is inadequate. In fact, some managers make natural mechanical aptitude a top priority when hiring MEs. In short, people who have no mechanical aptitude should probably look at working in fields other than mechanical engineering.

Now you have a basic understanding of mechanical engineering and the skills needed to perform well in the profession. Next, let's move into a section on the specific responsibilities of MEs.

Responsibilities

Mechanical engineers have a variety of different responsibilities where they apply their education, natural ability, and work experience. This is understandable based on the relatively high wages these individuals are paid in exchange for their services. They are expected to provide a return-on-investment, and that return-on-investment starts with responsibility.

The following are specific responsibilities of mechanical engineers:

Research and development

This is often assigned to mechanical engineers employed in higher education, but research and development (R&D) is also a focal point in some companies. In fact, organizations that are heavily involved in manufacturing often have an entire department devoted to R&D. However, regardless of the setting, R&D is something that MEs are usually responsible for at some point in their careers.

In higher education, R&D mechanical engineers often apply theoretical concepts in laboratory settings. This leads to new discoveries and innovation that drives technology in business. However, the downside is the fact that not all laboratory findings have real-world application.

In industry, MEs with R&D responsibilities usually come up with new ideas for machines and other mechanical systems. They start with an idea or concept and try to bring it from the laboratory to the production floor where it can be utilized to lower costs and raise efficiency.

Equipment and machinery design

This is one of the most common responsibilities of mechanical engineers. They design equipment and machinery with a variety of different factors in mind including quality, safety, cost, and efficiency. These factors are broken down as follows:

Quality

Quality refers to the specifications of products, equipment, or machinery. If ranges or tolerances are specified, then mechanical engineering personnel are responsible for meeting them.

Safety

This refers to safety of the personnel operating equipment or machinery. Various safeguards need to be put in place so operators do not get hurt. Without these safeguards in place, injuries can be very serious....and even fatal in some situations.

Cost

Cost refers to the money needed to get the equipment or machinery from the design phase to the shop floor. Essentially, it the labor and materials required to make the project successful. Cost also refers to the time and effort required to run and maintain the machine once it is operational.

Efficiency

Mechanical engineers need to get involved to make sure the equipment or machinery is meeting productivity expectations. Efficiency is very important in production-based facilities...and that is why MEs are employed in manufacturing plants.

Equipment and machinery modification

As noted above, efficiency is the responsibility of mechanical engineers. If equipment and machinery are not meeting performance expectations, then they need to be modified. Sometimes this can be done in process, and other time it means going back to the drawing board. Either way, MEs are responsible for performance modifications.

Process design

Some people assume that a lack of efficiency is due to the design of equipment or machinery. This is sometimes true, but it is not always an accurate assumption. The problem can be the process itself...and mechanical engineers are responsible for improving that process. They make changes to increase efficiency and lower costs based on observations and calculations.

Training

It should not come as a surprise that employees need to be trained on various operational and safety aspects of the equipment and machinery they operate. When this need arises, mechanical engineers are often the trainers because they were involved with that equipment and machinery from conception to implementation. They understand what needs to be done for safe and efficient operation, and they can answer any employee questions.

Project management

Most mechanical engineers are responsible for some type of project management. They oversee new production lines being implemented, procedures being changed, or expansions of workspaces. Regardless of the type of project, they are responsible for seeing it through to completion and this involves the management of people and processes. In this capacity, they also work as a liaison with outside contractors to ensure the project is finished in a timely and efficient manner.

Now you understand some of the major responsibilities of mechanical engineers. This leads us to discuss the surroundings where they perform their jobs...otherwise known as their work environment.

Work environment

This section discusses the conditions under which mechanical engineers perform their jobs. Work environment warrants discussion because MEs do not always work out of an office. In fact, some working conditions are far from that of an office.

The following are all environments where mechanical engineers work:

Manufacturers

This type of operation involves production lines that are used to assemble products. It is an excellent environment for the skills of mechanical engineers because equipment and machinery are crucial to the manufacturer's success. MEs are involved in every aspect of the production process, and the pace is often fast and furious. Manufacturing environments are action-packed and rewarding in terms of accomplishments…but they can also be stressful and lead to burnout.

Job Shops

This environment is similar to manufacturing due to the machinery involved, but there is typically no assembly line involved. Small numbers of custom made items are usually produced to pre-determined specifications. An order can go back and forth to various areas of the shop, and flexibility is more important than productivity.

Some mechanical engineers prefer job shops over manufacturers because employees are more skilled and typically more concerned about their jobs. They tend to treat their machines better because they take ownership of their jobs, and this means less repair or modification is needed. Jobs shops also tend to have less turnover than manufacturers, thereby reducing the need for operational and safety training that are sometimes conducted by MEs.

Laboratories

Laboratories are typically reserved for R&D and quality work performed by mechanical engineers. Laboratory MEs design new products and make sure machinery and equipment are meeting designated standards. Laboratories are generally lower pressure environments than production or job shop floors, and this is why some MEs prefer working in them. Laboratories are also desirable because they tend to be on the cutting edge of technology…but the downside is that they do not offer the excitement or challenges found in production facilities.

Offices

This refers to the traditional environment for many white-collar business people. Some mechanical engineers prefer doing their jobs from behind a desk because this works well for them. They miss out on the cutting edge technology offered in laboratories and the excitement of plant operations, but they are willing to sacrifice that for a nine-to-five work schedule where their main tools are a phone, computer, and calculator.

Some mechanical engineering work environments differ in location rather than type. This happens when MEs travel or take on temporary assignments, both of which are described below.

Travel

Mechanical engineers sometimes need to travel to locations where problems are occurring so they can resolve them. This is especially when MEs are employed by consulting firms, but it also applies to companies that operate facilities in multiple locations. MEs need to visit sites that need their services, and they stay as long as necessary....usually from one day to two weeks. Sometimes their services are needed on a periodic basis. For example, a mechanical engineer might need to travel to the bakery division of multi-facility food processor every two months to address a variety of issues.

Temporary assignments

Temporary assignments are similar to travel with the difference being the length of time at the location. In this situation, mechanical engineers are required to find housing for periods that can last several months to several years. Most MEs have a goal of getting back to their home base, but some prefer temporary assignments due to the change of scenery that prevents work environments from becoming stagnant. This change also provides new ideas and concepts that MEs can use for future problem-solving.

As you can see, work environments of mechanical engineers differ depending on their specific job. However, regardless of their position, they need to ready for change because it is going to happen. This leads us to the last section that looks at future changes that will take place in mechanical engineering.

Future

In general, the future looks bright for mechanical engineers. This is due to the fact that mechanical systems will always need to be designed, manufactured, tested, and maintained. ME skills will be needed for a variety of different applications due to the problems that need to be solved. However, as might be expected, MEs will face obstacles as they move forward. Some of the challenges they will encounter include:

- ### General

The following are general challenges that mechanical engineers will encounter:

Technology

Technology is important for mechanical engineers, but it will be even more important in the future. This is due to the fact that global completion will increase, and technological advances will play a critical role in which organizations are the most successful. MEs will need to understand, acquire, and implement technology in order to perform at optimal levels internationally. Astute MEs will realize that this technology comes from a wide variety of business applications that differ from those they find most comfortable.

Environment

Public perception of organizations today is often based on the impact those organizations have on the environment...and this is going to intensify in the future. Mechanical engineers will need to design and employ mechanical systems with greater respect for the natural world. Waste and pollutions will be minimized, and recycling will play a bigger role in decision making. In short, MEs will improve the image of their employers though environment responsibility.

Education

Mechanical engineers will need to continuously update their skills...regardless of their experience or educational background. Learning opportunities in the form of webinars and online courses will continue to grow, and MEs will be expected to capitalize on them. Employers will realize the value of education and they will gladly pay for these types of services.

Cost

Virtually every organization is impacted by money...and money will play a role in the future of mechanical engineering. Lower costs are necessary for lean manufacturing; and this means savings on research, design, operation, and maintenance of equipment and machinery will be more important than ever.

Diversity

As with most forms of engineering, mechanical engineering it is dominated by white males. This creates a homogeneous workforce that prevents some opportunities for growth. In a nutshell, gender and minorities will increase their presence as MEs, and businesses will benefit worldwide.

Specific

One major challenge that mechanical engineers will face in the future involves energy-related issues. Alternative sources of energy will need to be incorporated into the design of machines and equipment so their operation is energy efficient and cost effective. MEs will need to be conscious of the fact that traditional energy is not endless, and this will lead them to think about renewable sources for every aspect of their jobs. This is similar to what the automotive industry went through in the 1970s, except cost will be a larger factor than it was for car manufacturers.

Now you have an understanding of mechanical engineering. Let's use the same basic format to discuss the second branch of engineering known as electrical engineering.

Electrical Engineering

Introduction — 49
Education — 49
Natural ability — 50
Responsibilities — 51
Research and development — 51
Data analysis — 52
Computer-assisted design — 52
Equipment or machinery modification — 52
Training — 53
Project management — 53
Work environment — 53
Manufacturing — 53
Buildings and maintenance — 54
Laboratories — 54
Lighting — 54
Business owners — 55
Consulting — 55
Future — 55
General — 55
Specific — 57

Introduction

Electrical engineering is the branch of engineering that deals with the application and study of electronics and electricity. An electrical engineer (EE) performs a variety of different tasks at work, but his or her major job function is defined as:

The design, development, and testing of electrical components and systems.

The roots of electrical engineering can be traced back to the 17th century; and by 1800, a simple version of the electronic battery was developed. However, it was not until the 19th century, after the invention of the telephone and electric power, that electrical engineering became a field of study.

Modern day electrical engineering involves physics, math, computer programming, and electronics. People in this profession also understand thermodynamics, machine design, and systems engineering principles. They use this information to develop and analyze communication equipment, power generation machinery, and many different types of motors. They typically work in factories, laboratories, research facilities, and offices; and their work often has manufacturing or construction based application.

Top performing electrical engineers combine formal education with natural abilities for maximum effectiveness. These qualifications are defined in more detail as follows:

Education

Most electrical engineering jobs require a bachelor's degree in Electrical Engineering. This is obtainable from many different universities, and it usually takes four to five years to complete. In the United States, this degree is accredited by the *Accreditation Board for Engineering and Technology* (ABET). The ABET regulates course requirements for all colleges and universities to ensure adequate training and equal standards, and approved programs are listed on their website.

Much of the coursework in bachelor's degree programs focuses on math, science, and electronics. Math typically consists of higher level calculus, linear algebra, and differential equations; and science courses revolve around physics. Electronics courses explore problem-solving ...and other areas of engineering (mechanical, software, and power) are also part of the curriculum.

The following are types of courses taken be MEs In order to earn their bachelor's degrees:

- Calculus
- Chemistry
- Computer engineering
- Computer programming
- Digital logic
- Electrical circuits
- Linear algebra
- Mechanical engineering
- Physics
- Power engineering
- Probability
- Signals and systems
- Thermodynamics

In addition to the course work, most degree programs require the completion of a project to graduate. This usually takes place in a business rather than an educational facility, and it is designed to provide students with real-world experience and let them apply their problem-solving skills.

Natural ability

Electrical engineers' natural ability is often just as important as their education. As is the case with many engineers, EEs with the highest natural ability often turn out to be the most successful.

Natural ability includes:

Organization

Electrical engineers work can be complex when dealing with aspects of their jobs that involves circuitry or computer programs. For this reason, organizational skills are important. EEs without organizational skills often find themselves in confusing situations that result in guesswork...and this means they fall short of reaching goals and objectives.

Teams

Electrical engineers are often part of teams, so understanding how teams function and how to behave as a team member are important. Good team members understand that they cannot be in charge of every aspect because differing ideas and viewpoints will not be shared. They also realize that they cannot be "social loafers" who sit back and let others do the work. EEs who are effective team members are typically the most successful.

Communication

Most electrical engineers have valuable thoughts and ideas. They are trained to think and rationalize in order to solve problems. However, EEs are not always good communicators, and they sometimes have difficulty explaining and conveying their thinking. EEs who are naturally good communicators are able to transfer their thoughts and ideas to others, and this leads to the accomplishment of goals and objectives.

Trouble-shooting

Good maintenance people need to be more than mechanically inclined. If they cannot define the problem, then their repair skills are rendered ineffective. This same thinking applies to electrical engineers because EEs need to be able to troubleshoot problems in order to find solutions. This ability comes naturally to some people...and those people are usually the best EEs.

Now you have a basic understanding of electrical engineering and the abilities needed to perform well in the profession. Next, let's move into a section that focuses on specific EE responsibilities.

Responsibilities

Electrical engineers have a variety of job responsibilities that require them to find economical and timely solutions to problems with many different variables. They need to be able to apply theory, think rationally, perform multiple tasks, make decisions, and work with others. It is difficult to list every job duty of EEs, but their major responsibilities are as follows:

Research and development

R&D electrical engineers are often employed in higher education, but R&D is also conducted in organizations.

Automotive suppliers and manufacturers are good examples of companies that employ EEs in R&D. These companies often build multi-million dollar laboratories designed solely for EE research. In these labs, electrical engineers often use computer design software. They start with an idea or concept, build prototypes, prepare reports and other documentation, and present their findings to higher management.

In higher education, electrical engineers often apply theoretical concepts in laboratory settings. This leads to new discoveries and innovation that drives technology in business. However, the downside of laboratory R&D is the fact that not all findings have real-world application.

Data analysis

This ranks at the top of responsibilities for electrical engineers because they need to be capable of evaluating data received from components and systems in order to apply their knowledge and make decisions. Data analysis helps EEs design programs that create new products or solve problems encountered by employees in other areas of the organization. This makes everyone's job easier, and it saves time and money. That being said, people who do not possess data analysis skills probably should not be electrical engineers.

Computer-assisted design

Electrical engineers often work on projects. As part of the process, they typically use computer-assisted design (CAD) to visualize their ideas. These blueprints are important because they show flaws and prevent projects from being continued when problems show that they will not work. CAD allows EEs to efficiently put their ideas on paper or a screen, and it is a great tool for implementing specifications and adhering to other project requirements.

Equipment or machinery modification

This is an important responsibility for electrical engineers because they understand what machinery and equipment are capable of doing from an electrical standpoint. They observe operators in action and ask them questions pertaining to their jobs. After processing this information, they apply their knowledge and modify equipment or machinery to correct the problem.

Training

Some electrical engineering functions, such as CAD work or data analysis, are done strictly by EEs. However, this does not exempt them from training duties. Employees need to be trained on operational and safety aspects of the machinery they operate, and EEs are the best instructors because they can answer any questions related to the subject matter.

Project management

Electrical engineers work on many different projects. They test products, research cost-effective solutions, record and interpret data, and communicate their findings to customers and higher management. Regardless of the type of project EEs are assigned, they are responsible for seeing it through to completion…and this involves managing people and processes in an efficient manner.

Now you understand some of the responsibilities electrical engineers are charged with by their employees. Next, let's discuss the work environments where they perform their jobs.

Work environment

Electrical engineers work in a variety of different environments, and it would be difficult to list each and every one of those environments in the scope of this book. However, some of the more common types are listed below.

Manufacturing

This environment is typically production based where assembly of products takes place. Electrical engineers are involved in aspects of the production process that involve electricity or electrical components. Manufacturing environments are rewarding in terms of action and accomplishments, but they can also be stressful and lead to turnover.

Along the same lines, electrical engineers are also employed in power distribution environments. These environments are similar to manufacturing, but the focus is on distribution of power in controlled systems. Areas such as power flow, short circuit analysis, and system impact are examined by EEs as they research technology and design new products.

Buildings and maintenance

In this environment, electrical engineers have a multitude of responsibilities regarding the electrical aspects of building construction. These jobs require strong technical skills in addition to being able to multi-task and communicate with others. EEs comfortable with project management are often the best fit for this type of work situation.

Heating and cooling is another maintenance environment that employs electrical engineers. Heating, air conditioning, and cooling involve high voltage alternating current (HVAC) that requires the skills of EEs. They make sure indoor air quality and temperatures are maintained at acceptable levels by getting involved with all electrical aspects of the installation, operation, and maintenance of HVAC Systems.

Laboratories

Electrical engineers can find employment in a laboratory setting doing R&D work. In this environment, they typically design new products and make sure existing products meet electrical standards. Laboratory environments are generally lower pressure environments than production or job shop floors, and this is why some EEs prefer working in them. Laboratories are also desirable because they tend to be on the cutting edge of technology. However, the downside is that they do not offer the excitement or challenges found in production facilities.

Lighting

This is likely the most obvious type of work environment for electrical engineers. Most people understand that lighting involves electricity, so it makes sense that EEs are often involved. These engineers often work on commercial sites, overseeing all aspects of the lighting...from design to installation. This type of setting is good for EEs who are creative because projects often evolve and change in order to meet customer needs and code requirements.

Some electrical engineers are self-employed. In this capacity, they are business owners who sell their services to other organizations. EEs also work as consultants. Consultants differ from business owners offering EE services because they oversee the work rather than perform it. Below is a more detailed description of each.

Business owners

Electrical engineers who own businesses sell their skills to organizations that need them. They perform electrical engineering services for the organization that utilizes them, but they are paid as a company instead of an individual. This works well for some EEs because they are able to control their work schedule and they do not have to answer to a boss. However, the downside to owning their own business is the fact that they have to supply their own benefits and insurance.

Consulting

Electrical engineering consultants can work for themselves or a consulting agency. As noted above, the difference between a consultant and a business owner is the fact that consultants oversee the work being done rather than performing it themselves. They make suggestions for improving products and processes, instead of physically making the changes themselves. They often work with engineers employed by the company that hired them a consultant. In this sense, they tend to operate as project managers with the company engineers answering to them.

Base on the above, you can see that electrical engineers perform the functions of their jobs in a variety of different environments. This means they have options in their careers... and those options often provide motivation for being the best they can at work. This leads us to the last section that discusses the future of EEs.

Future

Electrical technology is ever-increasing, and the understanding of it is necessary for organizations to function efficiently. This takes a solid educational base and detailed work experience in the real world. Few people combine those skills, and that is why there will always be a demand for the talents of electrical engineers. However, there will be challenges involved as those engineers move forward. These challenges include:

General

The following are general challenges that electrical engineers will encounter:

First, it is important to note that the skills of electrical engineers will need to be continually updated. Those with the most education and experience will be employed in the best jobs. This means EEs will need to undergo constant training in order to keep abreast of new concepts and technology. However, it will also open the door for engineers of all nations to be employed by top-notch corporations...as long as they are willing to continuously improve. That being said, the following are general challenges that electrical engineers will encounter in the future:

Transportation

> Transportation is important for virtually everyone in the world. They might not own or lease their own vehicle, but they still need to be able to get to different places. The need for better and more energy efficient transportation will be a problem in the future, and electric vehicles will be a part of the solution. Electrical engineering technology will improve existing transportation by making it cleaner, safer, and more affordable. This will provide huge opportunities for EEs...but it will also pressure them to continually come up with new ideas and concepts.

Genetics

> Genetic engineering will need electrical engineers in order to progress. In fact, it will be almost completely reliant on the detection, analysis, and simulation skills of EEs. Electrical engineers will be at the forefront of altering genes for improving the biological capabilities of people and other living organisms. In the future, genetic engineering ranks as a major challenge for EEs.

Social responsibility

> Electrical engineers will need to make decisions with humans, ethics, and nature in mind. Profit will not be the only motive behind EE actions, and their main concern will be the society at large. Honest claims about research findings will be more important, environmental impact will have major significance, and consumer safety will be at the top of the list when designing new products. There are workable solutions to being socially responsible, but those solutions will come with many challenges.

Medical

Diagnostic tools used by professionals in the medical field will rely heavily on electrical engineering design and technology. This will highlight the significance of EEs, but it also means the health and safety of patients will become their responsibility. For example, heart pacemakers relying on electrical engineering technology will be able to save people's lives...but the malfunctioning of these units will have the potential to be deadly.

Diversity

Lack of diversity is a common challenge for all types of engineering, but it needs to be noted for EEs. In short, electrical engineering is dominated by white males. This creates a homogeneous workforce that prevents some opportunities for growth. In the future, gender and minorities will increase their presence as EEs...and businesses will benefit immensely.

Specific

One major challenge that electrical engineers will face is the need to acquire skills from other branches of engineering. Students majoring in electrical engineering will need to minor in mechanical engineering, civil engineering, or chemical engineering in order to meet the demands of global completion. Technical knowledge will rise in importance as EEs develop products and resolve problems in organizations all over the world. Some EEs are already experiencing this type of change, but it will become even more significant in the future.

Now you have an understanding of electrical engineering. Let's use the same basic format to discuss the third branch of engineering known as chemical engineering.

Chemical Engineering

Introduction	59
Education	59
Natural ability	60
Responsibilities	61
Safety	62
Product development	62
Evaluation	62
Design	62
Environment	63
Training	63
Work environment	63
Offices	63
Manufacturing	63
Laboratories	64
Worksites	64
Consulting	64
Travel	64
Future	65
General	65
Specific	66

Introduction

Chemical engineering is the branch of engineering that uses science and math to solve problems related to food, chemicals, pharmaceuticals, and other products. The job function of a chemical engineer (CHE) is defined as:

> *The conversion of chemicals, energy, and materials into useful products or processes.*

The first basic courses in chemical engineering were taught in the United States in the late 1800s. However, it was not until the mid-1900s that this branch of engineering gained recognition and was more clearly defined as a discipline.

Modern day chemical engineering involves chemistry, physics, microbiology, biochemistry, calculus, thermodynamics, and kinetics. CHE's knowledge of these subjects is used to solve a wide variety of problems associated with work-related conditions, processes, and procedures. These engineers typically work in factories, offices, and laboratories.

Top-notch chemical engineers combine formal education with natural abilities for maximum effectiveness. These qualifications are defined in more detail as follows:

Education

Most chemical engineering jobs require a bachelor's degree in Chemical Engineering. This is obtainable from many different universities, and it usually takes four to five years to complete. In the United States, this degree is accredited by the *Accreditation Board for Engineering and Technology* (ABET). The ABET regulates course requirements for all college and universities to ensure adequate training and equal standards, and approved programs are listed on their website.

The coursework in bachelor's degree programs focuses on science, math, and engineering related subjects. Science typically consists of chemistry, physics, and biology; math revolves around higher level calculus; and engineering related subjects include thermodynamics, process design, and biotechnology.

The following are types of courses taken be CHEs in order to earn their bachelor's degrees:

- Biology

- Calculus
- Chemistry
- Computer engineering
- Data analysis
- Fluid Mechanics
- Heat transfer
- Kinetics
- Linear algebra
- Organic chemistry
- Physics
- Product design
- Thermodynamics

In addition to the course work, most degree programs require the completion of a project to graduate. This usually takes place in a business rather than an educational facility, and it is designed to provide students with real-world experience that allows them to apply their problem-solving skills.

Natural ability

Chemical engineers' natural ability is often just as important as their education. As is the case with many engineers, CHEs with the highest natural ability often turn out to be the most successful.

Natural ability includes:

Problem-solving

The natural ability to solve problems is important for any engineer, but it is particularly important for chemical engineers because they are often addressing multiple problems at the same time. For example, CHEs addressing efficiency problems with processes in manufacturing hazardous chemicals also need to be concerned with worker safety and environmental impact. They must have the problem-solving skills necessary to correct the issue, safeguard employees' physical well-being, and prevent any type of damage to the environment.

Resourcefulness

Chemical engineers need to be innovative and creative when they apply old techniques to new problems. These types of situations require them to think "outside the box" and custom design solutions. Unfortunately, solutions are not always obvious, so CHEs need to be clever and original...and this is when their natural ability to be resourceful comes in handy.

Methodical

A major part of problem-solving is figuring out why something does not work as planned. This analysis requires a step-by-step process that eliminates some factors and brings others to the forefront. Questions need to be asked and trial-and-error often plays a big role. In short, systematic problem solving is important for CHEs...and those who possess the natural ability to be methodical find it very advantageous.

Logical

Chemical engineers must use sound reasoning when designing, analyzing, or troubleshooting situations. In other words, they need to take a logical approach to problem-solving. This might seem rather simple, but it can be challenging when multiple variables are often involved. Fortunately, the ability to apply logic in problem-solving can be learned...but those who naturally possess the trait will have a big head start.

Now you have a basic understanding of the natural abilities chemical engineers need to perform well in their profession. Next, let's move into a section on CHE's specific responsibilities.

Responsibilities

Chemical engineers perform a variety of different job tasks, and the responsibilities associated with those tasks can vary extensively from one engineer to another. For example, some CHEs work only with specific processes such as polymerization, while other focus on entire fields such as biomedical engineering. They are further divided within these areas because some CHEs do research while others work on manufacturing or process problems. Add this to the fact that CHEs work in many different industries (energy, food, plastics, paper, automotive, etc.), and it is rather obvious that listing every type of responsibility assigned to them would be very difficult. However, the following are some of their major duties:

Safety

Safety is a responsibility of many chemical engineers because they work in environments that manufacture, process, utilize, or handle hazardous chemicals. CHEs develop procedures for working with these chemicals based on their understanding of chemical reactions and the potential for injurious or lethal situations. If something is not safe, CHEs work to find alternative processes, raw materials, or chemistry to reduce the risk of injury. The importance of this responsibility should never be underestimated because it prevents humans from being harmed or, in extreme situations, killed.

Product development

Chemical engineers are often involved in product development. Interestingly, they sometimes find success by taking a backward approach. They are well trained for finding more efficient ways of doing things, so their road to product development often starts with defining better processes. They identify workable solutions to process problems and then work backward to develop unique products that are safe and cost-effective. This is not a new concept for scientists and, in reality, CHEs make up some of the best scientists in the world.

Evaluation

Solutions to problems in manufacturing are often found by chemical engineers. They troubleshoot by evaluating situations with a "think before act" mentality. They analyze equipment and processes first, and then they make decisions based on their education and experience. This evaluation procedure is similar to that used by other engineers...with one difference. That difference is the fact that CHEs always find solutions to problems with safety and environmental concerns in mind. If the end result has a safety violation or negative environmental impact, then CHEs work toward a better alternative.

Design

This refers to process design rather than product design. Process design is the centerpiece of chemical engineering because it extracts and utilizes all components in the field. Because of this, CHEs are often charged with developing new processes that resolve existing issues. They implement plans with procedures that coincide with equipment layout for maximum safety and efficiency. In terms of design, CHEs are usually the best employees that organizations have to offer.

Environment

As noted earlier, environmental concerns play a big role in chemical engineering. CHEs work to prevent environmental mishaps from occurring, and they also find solutions to environmental problems that already exist. They do this by implementing procedures that control pollution, reduce waste, and conserve energy. Mother nature never takes a backseat in a CHE's work, and this is what differentiates this field from other engineering types.

Training

Yes, training is again mentioned as a responsibility of a specific type of engineering. Chemical engineers usually conduct some type of training, and it is often geared toward safety. CHEs are responsible for the safety of personnel at all times, and they invest substantial time and effort into preventing work-related injuries. Their training efforts are important...especially for employees who work with dangerous or hazardous chemicals.

Now you understand some of the basic responsibilities of chemical engineers. Let's move on to a discussion on the work environments where they perform their jobs.

Work environment

Surprisingly, many chemical engineers work along with other white-collar workers in traditional office settings...although this is not always the case. The following are specific environments where CHEs find employment:

Offices

As noted above, chemical engineers frequently work in offices. They do this simply because they are able to work behind a desk and perform their job functions. They write policies, programs, and procedures based on their expertise; and their specifications are implemented by processing or manufacturing personnel. Training might be required, but sometimes even that can be done in an office setting.

Manufacturing

Sometimes chemical engineers find employment in manufacturing processes. On rare occasions, they act as supervisors who oversee production personnel, but they are usually hired to monitor processes. This process monitoring might be for efficiency reasons, but it is more often related to safety and environmental concerns. CHEs can work for just about any manufacturing plant, but they are typically found in chemical, food, clothing, and fuel production facilities.

Laboratories

Similar to other engineers, chemical engineers also work in laboratory settings. However, CHEs are probably the most likely type of engineer to find employment in this environment because a large part of their jobs involve research. They look for ways to improve products or processes, and much of this work can be done in laboratories. For example, they might be working on ways to improve the quality of food products...and this starts with laboratory testing and analysis. Laboratories also offer the opportunity to experiment with new technology that is not ready to be part of plant or production processes. Some CHEs prefer working with new technology over any other type of job function, and that a major reason why they chose to work in the field of chemical engineering.

Worksites

This refers to local worksites where chemical engineers find the need for their services. For example, many chemical spills require CHEs to spend hours, days, or weeks overseeing the cleanup. They understand the safety and environmental concerns involved, and they can prevent many other problems from occurring. In a situation involving a hazardous spill, CHEs are well worth the money spent for their expertise.

Consulting

Chemical engineers are also employed as consultants. In this role, they suggest ways to improve products or process while emphasizing the importance of safety and the environment. For example, they might be hired by a pharmaceutical company to find ways to make a particular drug safer for human consumption. Their suggestions might cost a substantial amount of upfront money, but they prevent problems such as lawsuits from occurring in the future. For this reason, their expertise is highly valued by management in organizations.

Travel

Sometimes travel is required for chemical engineers. This is unavoidable if, for example, they need to oversee safety concerns at a worksite that is not local. Obviously, the worksite cannot be brought to the CHEs...so they need to travel to it. This travel can last from one day to two weeks or longer...and repeat visits might be necessary.

Now you have a basic understanding of chemical engineering. Let's move forward to the next section that looks at the future of CHEs.

Future

Chemical engineers will be in demand, but they will likely have less overall opportunities than other types of engineers because they tend to be more specialized. However, the upside to this is the fact that CHEs who find employment will be paid handsomely because of their specialized skills.

Some of the future challenges chemical engineers will face include:

General

The following are general areas where chemical engineers will face challenges:

Medical

New medicines will be needed to treat cancer and virus-based diseases. Working with chemists and medical professionals, chemical engineers will be at the forefront of finding safe and effective drugs to fight these deadly illnesses. They will do this using their expertise to discover new materials and processes that change the current ways of thinking. The field of medicine offers many opportunities for CHEs, but those opportunities will be challenging because they will require change that takes people out of their comfort zones.

Biomedical

Chemical engineers will get involved with DNA, RNA, organs, and tissues in order to improve the longevity and quality of human lives. For example, they might conduct research to find ways to regenerate body organs, thereby reducing the need for donors. New products and procedures will be developed by

CHEs that put biomedical ideas into action. Thoughts that were once only a dream will become reality...as long as CHEs are willing to invest the necessary time and effort.

Ecological

The environment has been a concern for many years, and that concern will intensify in the future. Chemical engineers understand the chemistry involved with the earth, water, and air; and they will use their knowledge to reduce pollution, waste, and other threats to the natural world. Ecological issues will present problems, but CHEs will be able to provide the solutions.

Economical

Cost is an area where future chemical engineers will have an opportunity to shine. For example, they can improve the living standards for millions of people using low-cost energy solutions. Their research skills will lead to uses of solar energy that will be economically and environmentally beneficial. They will also work on new processes to harness wind and ocean currents for energy sources that have not been utilized to their potential. The challenge will be to produce the most amount of energy with the least amount of environmental impact at the lowest cost. This will be difficult, but it is achievable.

Specific

A major challenge that chemical engineers will face involves the climate. Climate change is a real threat, and future CHEs will need to address that threat using their expertise. This will require the implementation of intervention strategies designed to prevent further damage and reverse the harm that has already been done. It will also require the rebuilding of ecosystems starting with those that have had the most devastating impact. This challenge will be difficult, but it can be done...and it starts with CHEs.

Now that you have an understanding of chemical engineering, let's move on to the fourth and final branch of engineering known as civil engineering.

Civil Engineering

Introduction	68
Education	68
Natural ability	69
Responsibilities	70
Construction engineering	70
Geotechnical engineering	71
Structural engineering	71
Transportation engineering	71
Surveying	71
Planning	72
Managing	72
Reporting	72
Testing	72
Maintaining	72
Work environment	73
Government	73
Utilities	73
Non-residential buildings	73
Non-residential structures	74
Mobile	74
Future	74
General	74
Specific	75

Introduction

Civil engineering is the last major branch of engineering that will be discussed in this book. It deals with the overseeing of physical structures including buildings, roads, highways, bridges, tunnels, pipelines, and sewer systems. For simplification proposes, the job function of a civil engineer (CVE) is defined as follows:

> *The planning, design, analysis, and implementation of infrastructure in urban and rural areas.*

The roots of civil engineering can be traced back to the 1700s. At that time, military engineers were working on projects specifically designed for defense. Other engineers were working on non-military projects, and both types needed a way to be distinguished...so the term "civil engineer" was established. Interestingly, many of the functions performed by military engineers of the past have been taken over by civil engineers. This makes CVEs the most important of all engineers in terms of infrastructure development.

Present day civil engineering utilizes math, science, and other engineering disciplines for problem-solving. Math involves calculus, linear algebra, statistics, and differential equations; science revolves around physics, but geology and chemistry also play; and the other engineering disciplines consist mainly of structural, materials, and environmental. A CVE's work typically involves construction of non-residential buildings or other structural projects that focus on the transportation of people, solids, liquids, or gases.

The best CVEs combine education with natural abilities for effectiveness. These qualifications are defined in more detail as follows:

Education

Most civil engineering jobs require a bachelor's degree in Civil Engineering. This is obtainable from many different universities, and it usually takes four to five years to complete. In the United States, this degree is accredited by the *Accreditation Board for Engineering and Technology* (ABET). The ABET regulates course requirements for all colleges and universities to ensure adequate training and equal standards, and approved programs are listed on their website.

The following types of courses are typically taken be CVEs in order to earn their bachelor's degrees:

- Calculus
- Chemistry
- Computer-aided design (CAD)
- Computer engineering
- Differential equations
- Fluids and heat transfer
- Geology
- Hydraulics
- Linear algebra
- Physics
- Structural analysis
- Surveying
- Transportation engineering
- Thermodynamics

In addition to the course work, most civil engineering degree programs require the completion of a project to graduate. This usually takes place in a business rather than an educational facility, and it is designed to provide students with real-world experience and let them apply their problem-solving skills.

Natural ability

Civil engineers' natural ability is often just as important as their education, and those with the highest natural ability typically turn out to be the most successful.

Natural ability includes:

Communication

Civil engineers need to explain their ideas and plans to a variety of different people in work-related situations. Based on this, it is understandable that clarity is important for CVEs. They need to be able to clearly communicate with others in order to avoid the mishaps and problems that prevent projects from being successful. This natural ability is important for all engineers, but it is critical for CVEs to ensure problems are solved projects are properly completed.

Critical thinking

Civil engineers face a wide variety of problems that need workable solutions. In order to be successful, they need to gather and process information. They do this by combining known concepts with sound reasoning. This allows them to formulate conclusions while assessing the strengths and weaknesses of those conclusions. In short, CVEs need to properly evaluate situations...and critical thinking is the core of evaluation.

Decision making

Not surprisingly, civil engineers are often in charge of projects. They need to keep those projects moving forward while making sure applicable codes, rules, regulations, and standards are being upheld. This is accomplished by telling people the things that need to be done and instructing them on how to do those things. Based on this, it is understandable that decision making is a desirable trait for CVEs. Without the natural ability to make decisions, CVEs will not perform at levels expected of them in their roles as project managers.

Vision

This might be the most important natural ability for civil engineers. CVEs often oversee projects from conception to completion, and this requires them to visualize what needs to be done. Vision is something that cannot be obtained through formal education. It is a natural ability, and people who do not possess it should likely look at working in fields other than civil engineering.

Now you have a basic understanding of the education and abilities necessary for civil engineers. Next, let's move into a section on the specific responsibilities of CVEs.

Responsibilities

Before going on to specific responsibilities, it is important to note that civil engineers usually specialize in one of the following areas:

Construction engineering

These engineers deal with the planning, execution, and management of construction and other infrastructure such as tunnels, roads, buildings and utilities. These engineers combine the skills of civil engineers and construction supervisors so they can see projects through to completion.

Geotechnical engineering

These engineers are concerned with materials in the earth (dirt, rocks, minerals, etc.) and the construction that occurs on those materials. They make sure foundations are solid, redirect water flow, and develop retaining walls for specific applications. Essentially, civil engineers focus on structures, and geotechnical engineers focus on the support for those structures.

Structural engineering

These engineers calculate the stability and durability of structures such as building, bridges, dams, or sewer systems. They are skilled in structural design and understand the requirements necessary for safety. Their work is often conducted before any part of the structure is built, but their expertise is also required throughout the project.

Transportation engineering

These engineers apply science to the establishment and management of projects involving transportation. More specifically, they focus on the safety and convenience of transportation systems by overseeing the engineering aspects of construction. Major projects include the construction of streets, highways, railways, airports, streetcars, and airports.

The above specializations can overlap in some situations, but they need to be understood as divisions of civil engineering in order to get a clearer picture of CVE responsibilities. These responsibilities are as follows:

Surveying

This is likely the most well-known duty of civil engineers. In this capacity, they access and interpret land and geographically information from the job sites where they are working. More specifically, CVEs are often responsible for grades, elevations, and reference points necessary for guiding the building or construction process.

Planning

This responsibility involves the planning of systems and structures. Planning is necessary because standards need to be adhered to and requirements need to be met. It starts with a vision that typically utilizes computer-assisted design (CAD) to implement project specifications and indicate flaws that could lead to failure. Once the vision is found to be acceptable, the plan is able to move into the construction stage.

Managing

Civil engineers are often responsible for construction projects. This involves overseeing people and processes, and it means that CVEs need to manage. Without management, projects move in many different directions and goals fail to be achieved. CVEs often make the best project managers because they understand what needs to be accomplished and the most efficient path to those accomplishments. In short, their expertise is used for decision making that leads to objective achievement.

Reporting

Civil engineers are often assigned the reporting duties. For example, they need to conduct site investigations and report findings. They also need to research ergonomic, economic, and environmental concerns and report discoveries. Additionally, they need to examine proposals, bids, deeds, and leans to report discrepancies. These reports are important for any type of building or construction...and they are the responsibly of CVEs.

Testing

This is essentially a form of research and development, but it specifically involves the assessment of materials used in construction. Civil engineers understand which building materials are best for jobs based on their expertise, and they test those materials to make sure they meet specific requirements. In short, CVEs are assigned the responsibility of making sure materials are safe, meet code, and will do what they are designed to do.

Maintaining

All construction must be maintained. It might be years before a structure needs to be serviced or repaired, but it will happen at some point...and someone needs to be responsible. Civil engineers oversee maintenance projects such as the repair or replacement bridges, dams, highways, pipelines, tunnels, roads, sewers, and other infrastructure. This responsibility is important because it keeps construction effective, safe, efficient, and modern.

Now you understand some of the major responsibilities of civil engineers, so let's move forward to the next section that discusses their work environment.

Work environment

Work environments for civil engineers vary depending on their career choices. For example, a CVE who works for a city government will likely experience a vastly different environment than a CVE who builds highways in foreign nations. However, these individuals will share common job functions.

The following are some of the more common work environments for civil engineers:

Government

Civil engineers often find work in state, local, or federal government agencies. In this capacity, they work in offices or offsite, and their jobs can be somewhat diverse. For example, they might survey land for a certain period of time and then get involved with the building of government offices. This type of work environment is generally lower pressure than that in private industry, and that is why some CVEs prefer it.

Utilities

Utilities are chiefly made up of gas, water, and electricity. Civil engineers find employment in these environments in order to oversee projects involving sewers, pipelines, cables, and wires. They understand the materials being used and the impact these materials have on the earth. In short, utility companies rely on civil engineers for safety and efficiency.

Non-residential buildings

Office buildings, exhibit halls, and stadiums are all examples of this type of work environment. The work for non-residential buildings often starts outdoors with the foundation and finishes indoors with structural details involving doors, stairways, and windows. Civil engineers are needed for the design and construction stages because they understand buildings from a practical, environmental, and safety aspect.

Non-residential structures

Statues, monuments, and artistic creations are examples of non-residential structures that civil engineers are involved with building. This work environment can be outdoors or indoors depending on needs or specifications. Similar to non-residential buildings, CVEs are involved with the foundation and structural details. However, many of these structures are not meant to have people inside, so safety tends to be less of a concern.

Mobile

Mobile work environments require civil engineers to travel with projects unit they are completed. Examples include roads, highways, and railways. The upside to this type of work is the fact that CVEs always get a change of environment, but the downside is that projects can go on for long periods of time with many days spent on the road traveling.

As you can see, the work environments of civil engineers vary extensively from job to job. They can work indoors, outdoors, underground, or on the water. However, regardless of the environment, all CVEs perform similar job functions. Now let's move on to the future of this profession.

Future

Building and construction will likely never end, and because of this, there will be many opportunities for civil engineers in the future. However, CVEs will not be without challenges as they move forward. Some of these challenges include:

General

General areas where civil engineers will face challenges include:

Political

Like it or not, civil engineers will need to be more political in the future. They will become more involved with the environment and infrastructure, and this will force them to become more involved in the related issues. CVEs will be on committees that establish policies for environmental protection, encroachment, and public safety. They will need to become astutely aware of the legal aspects of building and construction, and this will be challenging for CVEs who want nothing to do with politics.

Safety

Safety is already important for civil engineers, but that importance will increase in the future. People will want to know that the buildings and structures they utilize are safe, and CVEs will provide that assurance. If CVEs fail to raise the bar on safety, then injuries and lawsuits will result.

Resilience

This refers to the resilience of the buildings constructed. Civil engineers will need to make buildings better and stronger so they can withstand natural disasters such as tsunamis, earthquakes, and tornadoes. Obviously, complete resilience will never be achievable, but the goal will be to constantly improve.

Congestion

This refers to traffic congestion. More vehicles will be on roads in the future, and civil engineers will be charged with making sure traffic does not get out of hand. They will need to design more efficient roads, highways, tunnels, and bridges so people can get where they need to be in a reasonable amount of time. This will take planning because space is not always available, but it can be done...and CVEs will be responsible for leading the way.

Specific

A major future challenge faced by civil engineers involves pollution. More specifically, CVEs will need to find ways to stop environmental pollution. They will need to utilize alternative sources of energy that are less damaging to the environment. They will also need to protect the earth that structures are built on and prevent groundwater from becoming contaminated. Humans often pollute environments in the name of progress...and CVEs will need to formulate methods for preventing some or all of that pollution.

Skilled Trades In Manufacturing
Short and Simple Explanation Series
Book 3

Louis Bevoc

Published by
NutriNiche System LLC

Louis Bevoc books...simple explanations of complex subjects

Introduction — 79
Electrical — 79
Description — 80
Education and training — 80
Example — 80
Plumbing — 81
Description — 81
Education and training — 81
Example — 82
Welding — 82
Description — 82
Education and training — 83
Example — 83
Heating and cooling — 83
Description — 84
Education and training — 84
Example — 85
Machinery — 85
Millwrights — 85
Machinists — 86
Carpentry — 87
Description — 87
Education and training — 88
Example — 88
Tool & Die — 89
Description — 89
Education and training — 90
Example — 90
Model making — 90

Description	90
Education and training	91
Example	91

Summary 91

Introduction

This is the third book in a series of short and simple explanations of professions. For every book, the profession is described along with a discussion on the required education and training. These books are written so people can inform and educate themselves on various professions without having to understand difficult language or complex terminology...which is the underlying philosophy of all Louis Bevoc books.

Since the 1960s, the trend for young people who graduate high school has been to get a college education. The thinking behind this is that a college degree yields a return-on-investment that cannot be achieved via other types of training or experience. The increase in high school grads choosing to go to college has prevented many of them from pursuing careers in skilled trades. Classrooms are preferred over apprenticeships, meaning students pay upfront for education that can be used later on rather than earning money while learning from on the job experience. Interestingly, skilled tradespeople often earn more money than college graduates, but the perception and physical work involved leads to many young people staying away from skilled trades as a career choice.

This book focuses on many different types of skilled trades in manufacturing including:

- Electrical
- Plumbing
- Welding
- Heating and cooler
- Machinery
- Carpentry
- Tool & Die
- Model makers

Each trade is described and exemplified using real-world application and the training and education requirements necessary for certification are discussed. This information helps readers understand the trade in layman's terms and allows them to see it in action. The goal is for readers to develop basic knowledge about all of the trades discussed, so let's get started.

Electrical

The following pertains to skilled tradespeople who are employed in manufacturing plants in an electrical capacity:

Description

Electrical skilled tradespeople are commonly known as electricians. These workers specialize in electrical wiring of plants, machines, equipment, and mobile devices such as floor jacks and forklifts. Their skills are in high demand for production facilities, and that is why they are one of the most popular types of trades in manufacturing.

Electricians have a wide variety of responsibilities in manufacturing plants, but they are often charged with installations, upgrades, and repairs of the plant, equipment, and machinery. In this role, they often function as project managers by overseeing all electrical aspects of a job. In short, they are directly involved in the labor and management aspects of many types of electrical work.

Education and training

Electricians are usually unionized employees who start as apprentices, work their way into a journeyman status, and then become master electricians. Typically, they are apprentices for three years or more, learning the trade on the job from more experienced electricians. During their apprenticeship, they earn money, but not as much money as journeymen electricians. Apprentice electricians are also required to log classroom hours. This combination of work experience and classroom education prepares them for the certification they need to be officially licensed as journeyman electricians. Without this certification, they cannot perform many of the job functions that manufacturing plants require.

Master electricians are the highest level of the trade. To become a master, journeymen electricians need about ten years of on-the-job experience and also must pass a written exam. This is not an easy task because it requires working knowledge of the National Electric Code (NEC). The NEC is quite extensive and encompasses a wealth of information, but electricians who understand it truly are prepared to take on the challenges found in manufacturing facilities.

Example

Hector is a master electrician at a paper mill. The plant manager decides to put an addition on the plant for greater production capacity, and he needs all electrical wiring in the addition to meet local code. Hector understands the code, and he oversees all wiring so that it is compliant. This allows the plant manager to focus on production and quality related issues rather than worry about the addition passing the electrical inspection of local authorities.

Plumbing

The following pertains to skilled tradespeople who are employed in manufacturing plants in a plumbing capacity:

Description

Plumbing skilled tradespeople are typically known as plumbers. They can also be pipefitters, steamfitters, and boilermakers...but these trades sometimes have skills other than or in addition to plumbing. Pipefitters, steamfitters, and boilermakers often work as plumbers, but plumbers focus on low-pressure piping systems that convey water while pipefitters, steamfitters, and boilermakers work with higher pressure systems for moving steam, chemicals, and fuel. However, pipefitters, steamfitters, and boilermakers might not have the same plumbing skills as plumbers because their jobs are not specifically geared toward plumbing problems.

Plumbers are valuable to manufactures because water, steam, chemicals, and fuel must be moved via lines of piping in virtually every plant. Plumbers keep these lines open; thereby avoiding hindrance of production processes. They also work on sanitary lines, potable water structures, and sewerage systems. Backed up drains in production facilities can lead to many different problems including plant, equipment, and machinery damage. Plugged drains have even been known to bring manufacturers to a complete halt, so it is rather obvious that plumbers are "worth their weight in gold" in these situations.

Plumbers handle small and big problems in manufacturing plants. In the same day, they might fix the faucet of a sink in the employee break room and then move on to repair a broken water main that has the potential to cease plant operations. Other responsibilities include reading drawings, relocating lines, installing apparatus, threading and soldering pipes, testing for leaks, working on drainage systems, and adhering to sanitary requirements. They must also be aware of safety standards, building regulations, and legal regulations when working in manufacturing facilities.

Education and training

It takes many years to become a knowledgeable plumber because the ability to troubleshoot problems and develop cost-effective solutions is essential. Training for this trade is best done on the job and, although there is no federal licensing available, state and local governments often require plumbers to be licensed for the work they perform. Unfortunately, some people believe plumbers are not governed by any rules, but this is simply not true.

Example

Edwardo is a plumber at a meat processing facility. There is a drain backup in the smokehouse area, and production has been shut down until the clog is removed and the drain flows freely. Edwardo understands that the clog is east of the drain, not west of it, because he understands the flow of the sewage lines running underneath the plant floor. He wastes little time as he runs his "snake" into the pipe 50 feet east of the drain. The clog is removed, and the drain begins to function properly due to Edwardo's knowledge of the sewer pipe system running under the plant.

Welding

The following pertains to skilled tradespeople who are employed in manufacturing plants in a welding capacity:

Description

Welding skilled tradespeople are typically known as welders. They use welding machinery to fuse materials together for bonding purposes. Most people associate welding with metal, but other materials welded include polymers and plastics.

Welders often have a sub-skill within their trade. For example, a metal welder might focus on stainless steel due to the challenges involved with this type of welding. Stainless steel has specific properties that vary from other metals, and many stainless welds are required to be smooth for use in food or pharmaceutical applications. Aluminum welders also possess a sub-skill because the metal is soft and requires skill not to damage. The low melting point and high thermal conductivity of aluminum can lead to burn through of the metal unless the welder has the necessary knowledge.

Welders are valuable to manufacturers because their skills are needed for the wide variety of machines necessary for production. They are capable of fabrication or repair, and their work is critical for systems and processes to operate smoothly and effectively. They are always needed in production geared facilities because machinery is necessary to generate a high volume of products, and high volume is the key to profitability for many manufacturers.

Welders are often charged with installing equipment and machinery. They are typically responsible for the design of the projects that involve them; thereby requiring them to have a practical working knowledge of physics and mathematics. They are also responsible for making sure safety and sanitary standards are adhered to, and they need to do their jobs with limited input from others. Some welders work on teams in manufacturing plants while others perform their jobs by themselves, but it is essential that they are all able to accomplish their goals and objectives without direct supervision.

Education and training

Training for welders can be done via a postsecondary associate degree, educational certificate, or apprenticeship that involves several months of on the job training under the supervision of an established welding mentor. Associate degrees in welding are offered by many accredited universities while educational certificates are offered by schools that specialize in welding. However, many welders prefer to learn the trade under the watchful eye of a skilled welder, and this is enough to get them jobs in many manufacturing plants.

Example

Devon is a welder at automotive parts manufacturer. The company is revamping one of its production lines so it can handle a larger volume of products, and Devon is the lead welder on the project. She has two other welders working under her, and her team needs to weld a series of conveyors so they are able to handle the increased flow of product. She coordinates every aspect of the welding, including assigning the other team members jobs and setting deadlines for completion. In this role, Devon is a welder and a lead worker which earns her a ten percent higher pay rate than any other welder at the plant.

Heating and cooling

The following pertains to skilled tradespeople who are employed in manufacturing plants in a heating and cooling capacity:

Description

These individuals are often employed as technologists for environmental comfort in buildings and vehicles. They work on heating, ventilation, and air conditioning systems (HVAC) in people's homes and cars. In this capacity, they are usually directly hired by homeowners or car owners to do specialized work involving repair or service.

In manufacturing, heating and cooling skilled tradespeople typically work on furnaces, refrigeration systems, and air conditioners. Their service might be as simple as adding Freon to an air conditioner or it might as complex as installing an entirely new refrigeration system.

Similar to welders, heating and cooling tradespeople often have a specific skill. One person's skill might be heating while another's is cooling, and some specialize in indoor work while others focus on outdoor jobs.

Heating and cooling tradespeople handle any job related to heating and cooling in manufacturing. They might work on a furnace for the offices or they might install a refrigeration system for a 1,000,000 square foot plant. They designate refrigerant types, determine capacities (tonnage), create designs, and oversee projects. They need to understand the basic principles of mechanical engineering, thermodynamics, heat transfer, and refrigeration. They often work in teams, but it is essential that they are able to work alone as many jobs do not require multiple technicians.

Education and training

Training for heating and cooling people is done in specialized schools and on the job under the supervision of an established tradesperson. Certificates can vary, but they often are as follows:

- *Type I Certificate* - For the installation, repair, and service of most small appliances in homes or offices (window air conditioners, refrigerators, freezer, vending machines, etc.).

- *Type II Certificate* - For the installation, repair, and service of high-pressure refrigerant equipment (heat pumps, central air conditioners, commercial refrigeration, etc.).

- *Type III Certificate* - For the installation, repair, and service of low-pressure refrigerant equipment in industry.

- *Type IV Certificate (Universal)* - For the installation, repair, service, and disposition of equipment containing any type of refrigerant.

Example

Rodriquez is employed at a heating and cooling company that services a metal casting foundry. Heating is a critical aspect of the processes at the foundry, and Rodriquez specialty is heating equipment. He visits the plant every Wednesday to perform routine service work, and he is on call 24/7 for equipment malfunctions and breakdowns.

Most of the jobs at the foundry can be handled by Rodriquez without any help, but he has two other technicians who are familiar with the foundry's process if they are needed for a team effort. The foundry understands the importance of having Rodriquez available at a moment's notice, so they pay a monthly retainer fee for his services.

Machinery

For simplification purposes, this book breaks down machinery tradespeople into two distinct subcategories known as millwrights and machinists. In manufacturing, millwrights fabricate, install, maintain, and repair machinery while machinists operate those machines to make products. The following examines both subcategories of machinery tradespeople who are employed in manufacturing:

Millwrights

Description

Millwrights typically perform work on machinery in manufacturing plants. This machinery is usually stationary (such as presses and mold makers) rather than mobile (such as floor jacks and forklifts), and the work involves installation, service, moving, or repair.

Millwrights' jobs are usually less specific than some of the other tradespeople, so they need to be quite versatile. They must read service manuals and schematics, implement preventive and predictive maintenance, align mechanical equipment, operate rigging equipment, and tack weld on a moment's notice. They must also understand the basics of electronics, pneumatics, and hydraulics and be able to work without direct supervision. Last, but certainly not least, millwrights sometimes need to perform work in high places, so they should not be afraid of height.

Education and training

Schools offer training and certificates for millwrights, but this type of tradesperson also learns under the watchful eyes of an experienced journeyman. Formal apprentice programs that last several years are often required by unions and contractors before a person earns the rank of a journeyman. Once that rank is achieved, millwrights are authorized to work without direct supervision.

Example

Terrance is a millwright in an automotive assembly plant. He and four other millwrights have been assigned the responsibility of installing a stamping press on the production floor. His specific job on this project is to make sure the machine is aligned correctly so it functions properly. He does not necessarily need the help of the other millwrights, but he can use their help if they are needed.

Terrance does not have any major weaknesses as a millwright, and he is capable of performing every type of task that falls under his job function. His experience and knowledge make him a valuable asset to the automotive assembly plant for a wide variety of different projects.

Machinists

Description

Similar to millwrights, machinist positions are often defined by unions. However, their major job function is to operate machinery in manufacturing plants. This machinery is usually stationary (such as presses and mold makers) rather than mobile (such as floor jacks and forklifts).

Machinists' jobs are often very specific because they work on the same machines for their entire shift. They must understand their machines from a technical and functional aspect because fine tuning is often necessary. They typically have direct supervisors, but they are expected to handle everyday tasks without management assistance. Although not required, machinists should also understand the basics of electronics, pneumatics, and hydraulics.

Education and training

Schools offer training and certificates for machinists but the most common way to learn the critical aspects of the job is to work under the guidance of a journeyman. This learning is important for many reasons including the fact that machine parts are often manufactured to high tolerances that require expertise. Add to this the fact that these parts are sometimes produced in high volume, and it is relatively easy to see why training is important.

Example

Matilda works as a machinist at a plastics factory. Her job is to make lids for six-ounce and twelve-ounce plastic containers. Precision is necessary for this job because any lid with a tolerance exceeding .005 inches will not snap shut properly on the container. She produces about 8000 lids during her eight-hour shift, so mistakes could result in a lot of scrap material.

Matilda trained at her job for five years before she was allowed to make the lids without direct supervision, and now she has a skill that is very valuable to the plastics manufacturer. Her attention to detail is critical because it could affect the profitability of her organization, but she is trusted because she is an experienced machinist.

Carpentry

The following pertains to skilled tradespeople who are employed in manufacturing plants in a carpentry capacity:

Description

It is difficult to say how long carpentry has been around since people have worked with wood long before it was thought to be a trade. Their work went virtually unnoticed because it was considered a part of the process that led to the completion of projects, and most men had some type of woodworking skills in order to survive. However, over time, those woodworking skills evolved into a trade with its own recognition, commonly known today as carpentry.

Part of the evolution of carpentry is due to technological advances. Over the years, humans learned how to mass manufacture tools so workers doing carpentry work could focus on the jobs they were doing rather than creating the tools necessary to properly do those jobs. Today, computerized devices have made the planning, setup, and installation processes of jobs faster and more efficient; thereby allowing carpenters to take on multiple projects at the same time and finish them faster than they ever could do in the past.

In manufacturing operations, these skilled tradespeople set up and maintain building structures. The work with many different materials including metal, linoleum, marble, fiberglass, plastic, concrete, and wood, and they use those materials for flooring, framing, and creation of storage areas. Often times, they perform their work in virgin or untouched territory; thereby making them "pioneers" at their job sites. In other words, carpenters are typically the first tradespeople on the job, doing rough work that others will complete. They do not need to stay around until the job is finished, but those who are employed by manufacturers often check in periodically to see how the work is going and answer questions. This makes sense because their job often requires them to work in many different areas of the same plant so they are normally within walking distance of their completed projects.

In short, carpentry is a well-known trade that has evolved since its humble beginnings. Manufacturers understand the value of these tradespeople and utilized their knowledge and skills on a regular basis.

Education and training

Not surprisingly, training is required before carpenters are certified in their trade. This training involves book learning and on the job learning. First, aspiring carpenters take pre-apprentice courses in union approved college programs. After completing these programs, they begin their journey as apprentices under the watchful eyes of mentors who have practiced the trade and understand it well. After four or five years, they become journeymen, and further down the line they can take a test to become masters. Obviously, there is more involved with becoming a master carpenter than is shown in this paragraph, but it shows the basic steps involved.

Example

Bertram is a journeyman carpenter at a computer hardware assembling facility. He has been charged with the task of putting in a new marble floor in the foyer at the entrance of the plant offices. He uses a computer to map out how he will do the work and how many floor tiles will be needed to complete the project. After he knows what he needs to do and the number of floor tiles required, he purchases the tiles via an online supplier. The tiles are delivered to the plant, and Bertram begins installing them. After the installation, he leaves the job ready for the drywall people to do their work.

Tool & Die

The following pertains to skilled tradespeople who are employed in manufacturing plants in a tool and die makers:

Description

Tool and die makers are unique tradespeople because they are usually only employed in manufacturing plants. They are machinists who make a variety of molds, dies, gauges and machine tools; most of which are used solely in production type operations.

Tool and die makers typically do not work in production areas. Instead, they are based out of machining shops or tool rooms so they have their own space, tools, and machinery to properly do their jobs. They can work alone or on teams that usually include engineers because engineers have many things in common with tool and die makers. For example, they both use math to resolve problems. Additionally, they both need a mechanical background in order to understand how machinery works. Last, but certainly not least, they both must have an understanding of science-based technology so they can apply the principles while trouble-shooting various manufacturing issues.

A major change tool and die tradespeople have undergone in recent years is they are now required to understand computer technology (it is no longer an option). For example, computer numerical control (CNC) is routinely referenced and utilized because it allows for the automation of machine tools via pre-programmed commands. This technological process is much faster, accurate, and efficient when compared to the wheels or levers that were hand-operated by employees in the past. Additionally, computer-aided design (CAD) allows for the creation, modification, and optimization of designs in a faster and more efficient manner.

In short, tool and die makers are needed in manufacturing, but they need to keep up with modern technology. Their worth is based on their ability to add value to production processes and, without technology, that value is limited.

Education and training

Most tool and die tradespeople undergo an apprentice under an experienced mentor. After about five years, they have enough experience to obtain the rank of journeyman and practice their trade without supervision. In other words, they are said to have mastered the necessary skills to properly do their jobs.

Interestingly, some tool and die makers graduate from accredited colleges and work thousands of hours to achieve journeyman status. This is by no means a shortcut, it is simply a way to become a tool and die tradesperson while earning a college degree.

Example

Donte is a tool and die maker employed at a small forklift assembly plant. One of his responsibilities involves designing dies for rollstock machinery that packages the paperwork and small parts that come with every forklift purchased by customers. Donte has a manufacturing engineer complete a computerized design of the dies along with the size and material requirements. He uses this design to create a 6-inch x 6-inch x 2-inch prototype using cast aluminum material. This prototype is tested on the rollstock machine by quality personnel and found to work properly. The prototype is returned to Donte and he creates three more identical dies (one for each rollstock pocket) for use by production employees.

Model making

Description

Model making tradespeople are assigned the responsibility of creating miniature representations of designs, ideas, and concepts. They create prototypes and mock-ups that can be critiqued before the actual construction or production begins. This critique is typically limited to the physical traits of the product or structure but it is also used for eye appeal; thereby making it valuable for sales and marketing personnel.

Model makers employed in manufacturing work in design shops, research facilities, and production areas. They need to know how to use a wide variety of equipment and machinery including lathes, saws, welders, and mills. Materials used include clay, wood, cement, metal, and plastic, and the overall process can get very detailed. These details include the finish and trim work because the end result needs to be representative of the product or structure designed for creation.

In short, model makers must have patience and the ability to pay attention to detail. Without these two traits, prototypes and mock-ups become rushed and the finished work does not appear professional. Professional appearance is important because, as with other jobs requiring manual labor, model makers are slowly being replaced by technological advances that are capable of saving money and time.

Education and training

As is the case with most trades, model makers need to go through some type of apprenticeship to achieve journeyman status. There is no specifically designed training program for this apprenticeship, but it should include practical design and model making. College coursework is typically not required even though it is available (although this requirement can vary depending on the employer). Quite simply, manufacturers prefer real-world experience over a certificate or degree. However, if the educational route is chosen, then creative courses, such as woodworking and ceramics, should be a major part of the curriculum along with model making and computer-aided design.

Example

Francine is a model maker for an architectural firm. She uses a blueprint developed by an architect to make a model of a building that will be used for visionary and structural purposes. The model is made out of brick and metal; thereby making it similar to the actual structure. After the building is created, it is critiqued by everyone at the firm involved in the building process, and modifications are made as deemed necessary. In the end, Francine's model is very representative of the real building, and it is approved by multiple people before any work is done on the actual structure.

Summary

Tradespeople are employed in manufacturing plants all over the world, and their contributions to the efficient operation of those plants cannot be underestimated. They work on processes ranging from simple to complex, and those processes produce the products necessary for organizational growth and prosperity.

This book explores skilled trades in manufacturing including electrical, plumbing, welding, heating and cooling, machinery, carpentry, tool & die, and model making. Each trade is described and the education and training necessary for certification or licensing are discussed. Examples are provided for further clarification, and the text is written for easy understanding at all reader levels.

Congratulations! You now understand more about skilled trades...essential jobs in every manufacturing facility.

www.ingramcontent.com/pod-product-compliance
Lightning Source LLC
Chambersburg PA
CBHW021454210526
45463CB00002B/770